STATISTICS
An introduction

Alan Graham

First published in Great Britain in 1988 by Hodder Education. An Hachette UK company.

This edition published 2017 by John Murray Learning.

Previously published as *Teach Yourself Statistics* and *Understand Statistics: Teach Yourself*.

British Library Cataloguing in Publication Data: a catalogue record for this title is available from the British Library.

Library of Congress Catalog Card Number: on file.

Paperback ISBN 978 1 47365 200 2

eBook 978 1 47365 201 9

1

Typeset by Cenveo Publisher Services.

Printed and bound in Great Britain by CPI Group (UK) Ltd., Croydon, CR0 4YY.

John Murray Learning policy is to use papers that are natural, renewable and recyclable products and made from wood grown in sustainable forests. The logging and manufacturing processes are expected to conform to the environmental regulations of the country of origin.

Carmelite House

50 Victoria Embankment

London EC4Y 0DZ

www.hodder.co.uk

Also available in ebook

Contents

Acknowledgements

Many thanks to Irene Dale, Tracy Johns, Ian Litton and Kevin McConway for their help and support.

Meet the author

Welcome to *Statistics: An introduction*!

Hello, my name is Alan; I've been a lecturer in Mathematics Education at the Open University for more than 30 years. My particular interest is statistics and I've written a number of books and Open University course units in this field. I've given numerous workshops and lectures to teachers on a variety of themes, including calculators, spreadsheets, practical mathematics and statistics.

Most statistics textbooks are too difficult to understand. This is not the fault of the people who read them but is down to the unrealistic expectations of the people who write them. Most learners starting statistics are, unsurprisingly, not very confident with handling numbers or formulas, so why pretend they should be?

This book is my attempt to present the basic ideas of statistics clearly and simply. It does not provide a comprehensive guide to statistical techniques – if you want that, you will need a different sort of book. Instead, the focus is on understanding the key concepts and principles of the subject. Increasingly, much of the technical side of statistical work – performing laborious calculations, applying complicated formulas and looking up values in statistical tables – is these days handled by computers and calculators. What really matters is to understand the principles of what you are doing and be able to spot other people's statistical jiggery-pokery when you come across it.

I believe that the ideas in this book will provide you with a bedrock understanding of what statistics is all about. I have enjoyed writing it. I hope that you will enjoy reading it.

Alan Graham, 2017

Introduction

For most people, the word statistics conjures up working with *numbers* and doing *calculations*. Well, accuracy in calculation is certainly important, but these days machines can take care of this side of things. With cheap but powerful calculators and computers around, the job of humans like you and me is to have a clear sense of what you want to do with your data so that you can set the machine to do the correct calculation, and then interpret the answer sensibly. Actually, it is the thinking and understanding that are key here, rather than the mindless number-crunching.

There are two main aims of this book. The first is to make you more confident about handling statistical information (percentages, tables, charts and so on). With this understanding, not only will you be able to grasp what is being said when figures are used, but you should also be better able to protect yourself when scam-merchants try to take you for a ride! A second aim of the book is to give you an understanding of some of the big themes of the subject.

So what *are* the 'big ideas' of statistics? At the most basic level, they include collecting data (including questionnaire design), summarizing data (including calculating averages) and plotting data in a graph or chart. At a more advanced level, they include sampling, regression and correlation, and carrying out a statistical test of significance (see the diagram below).

Carrying out a statistical investigation

Imagine that you were to tell a student of music that they were required to spend years practising scales and arpeggios but would never get to play a tune. Next try telling a student of soccer that they must devote months to learning how to head, trap and kick a ball but never get to play a game. As you can imagine, both of these students would waste no time in showing their displeasure. Yet this is often what happens when students learn subjects like mathematics and statistics. You can easily go through an entire statistics course learning how to work out averages, plot charts and graphs, calculate coefficients of correlation and so on, but never get to use them in a meaningful investigation. One of the challenges of this book is that while you are learning each new statistical skill you should try to relate it to the bigger picture of how it might be used for real in a simple statistical investigation.

In many textbooks on statistics, the main stages involved in carrying out a statistical investigation are presented in three steps, as set out below:

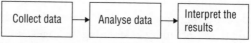

A statistical investigation with only three stages

However, my concern with this picture is that there is no indication of where these data came from in the first place or why they were ever collected. Without some sense of purpose and direction, statistical activities can easily fall flat on their face, leaving students none the wiser about the power of the statistical ideas they had been using and where they fit into a wider picture.

I would suggest that the vital missing box is located just before the 'collect data' stage and so I suggest adding a fourth stage at the beginning called 'Pose the question'. By starting with a clear, well-formulated question that you really want to investigate, the remaining stages follow naturally and your use of statistical skills and techniques can have a sense of purpose.

So the revised diagram looks like this:

The four stages of a statistical investigation

These four stages of a statistical investigation can be summarized by the letters PCAI. These are set out in the table below.

Stage 1	P	Pose the question
Stage 2	C	Collect the data (collecting and recording)
Stage 3	A	Analyse the data (summarizing and plotting)
Stage 4	I	Interpret the results

In practice, of course, investigations rarely follow the neat linear path suggested by these sorts of diagrams. It may be more helpful to see the model of statistical investigation as a cycle which you may need to travel around several times before reaching a satisfactory conclusion. This is shown below, with the important, and often neglected, first stage, Pose the question, highlighted.

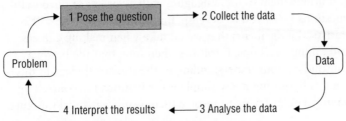

The four stages of a statistical investigation as a cycle

When you are tackling a statistical investigation yourself, this PCAI cycle can be a very useful aid to your planning. In particular, an awareness of the four PCAI headings can:

▶ help you to think through what needs to be done and when;

▶ provide criteria by which you can judge the quality of your outcomes;

▶ offer a useful set of headings for a write-up or oral presentation of your work;

- help you to generate a sense of what it means to be constructively critical of your methodology, always looking for alternative interpretations of the data;

- help you to focus attention on the links between successive stages of the cycle and use these as the basis of asking key questions about what you have done and why (for example, the P–C link prompts the question, 'Do the data being collected help to answer the question posed?', and so on); and

- encourage you to see the bigger picture of what statistics is really all about and so give a sense of purpose and direction to your statistical work.

Models such as PCAI are familiar enough in educational circles. However, for me this approach is not just about trying to place statistical work in a practical everyday context. What makes it special as a model for learning about, and learning to use, statistical ideas is the 'P' stage at the beginning. The posing of a question is what gives a purpose to the whole enterprise and it is a constant reminder to you of what you are doing, and why. It is the key that reveals the need for statistical ideas to be used and brings them to life. Engaging with a variety of purposeful questions helps you to develop an intuition about the sort of statistical judgements that we all take when making choices and taking decisions. It follows, therefore, that the best way to consolidate your understanding of the statistical ideas in this book is by posing a few simple investigations for yourself and having a go at trying to resolve them using statistical thinking.

How to succeed at studying this book

To end this section, here are four useful tips to help you to make your statistical journey through this book an exciting and successful one:

- go at your own speed and be prepared to re-read sections if necessary;

- ▶ where possible, relate the ideas to your own personal experiences;

- ▶ try to learn actively (this means actually doing the exercises, rather than saying vaguely, 'Oh yes, I'm sure I can do that'); and

- ▶ be prepared to explore and experiment with ideas, particularly by taking every opportunity to exploit technologies such as the calculator and computer.

1

Introducing statistics

In this chapter you will learn:

▶ *why learning about statistics might be a worthwhile thing to do*

▶ *about some everyday situations where having a better statistical understanding might be useful to you*

▶ *what sort of questions might be seen as 'statistical' questions*

▶ *whether it might help to use a calculator.*

Some good reasons to learn about statistics

Statistics are all around us – in newspapers, on television, in general conversation with friends. Here are a few typical examples of headlines, graphs and advertisements that you might easily find in a daily newspaper or online.

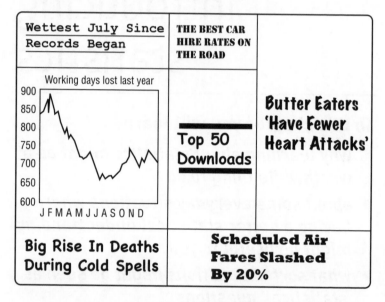

Although it may not yet be immediately obvious to you, each of these examples embodies an important idea in statistics. They suggest just a few of the choices that we make all the time, with or without an understanding of statistical ideas. An underlying aim of this book is to suggest to you that statistics isn't just something you learn about from a textbook, but it can actually change the way you see the world and inform your judgements. There are a number of more specific reasons why it makes sense to have a basic grasp of statistics. Below are listed just some of them.

▶ As these examples suggest, much of the information we have to process in our different life roles – at home, as consumers, at work, in the community and as citizens in a wider economic and political setting – comes in the form of numbers, graphs and charts. A statistical awareness helps us to understand the forces acting on us and to process the information that confronts us, particularly when it is being used in a deliberately misleading way.

▶ Many of the important decisions that we have to make involve handling numbers and weighing up risks. A knowledge of statistics and probability won't guarantee that you always make the *correct* decision, but at least your decisions should be better informed.

▶ From archaeology to zoology, almost every subject covered at school, college or in the media has become increasingly quantitative, so it is becoming ever harder to hide from the need to have a basic understanding of statistics. This is particularly true of business.

▶ Statistics can be enjoyable – see how you feel about this idea by the time you have finished this book!

Statistics in everyday life

It seems that there is no escape from statistics! Increasingly, a statistical awareness is an essential requirement of the workplace. For the majority of people, a day in work will include some time spent processing and interpreting information in the form of numbers, percentages, tables, graphs or diagrams.

But even outside the world of work, statistical and mathematical skills are an essential part of being a functioning adult, whether you are listening to the radio, watching television, reading a newspaper or magazine, cooking a meal, following an instruction manual, assembling furniture, checking train times, ordering concert tickets, ordering weekly food shop, filing tax return online … The list goes on and on.

When engaging in these sorts of everyday events, you are probably never fully conscious of the statistical thinking that you are using. But sometimes it is valuable to make your thinking explicit, if only to be aware of your own strengths and weaknesses and discover where you might need to brush up your skills and understanding. Below are just two particular areas where most of us could do with a bit of a brush up: handling very small and very large numbers and evaluating risk.

SMALL AND BIG NUMBERS

For most people, the sort of information that they handle on a day-to-day basis tends to involve whole numbers in the range 0 to 1000. For example, in a typical week you will certainly bump into the numbers 12 (the hours on a clock face), 30 (speed restrictions in miles per hour), 180 (playing darts or setting the oven temperature) and so on. And if you are a sports fan, numbers in the form of scores, times and distances will come at you in a constant stream. However, our exposure to very small numbers such as 9.1094×10^{-31} (the mass of an electron at rest in kg) or big numbers such as 299 792 (the speed of light in km per second) is much rarer. As a result, most people are not good at handling these sorts of numbers with confidence, and indeed human intuition is often flawed in situations where very small or very large numbers are involved. For example, the estimates of the number of people attending a public demonstration, football match or concert will vary enormously from one person to another. If you want to improve your number processing skills, try thinking up, in an idle moment, some questions to ponder similar to those shown below. Then try resolving them using rough, common-sense estimates and simple calculations (a calculator will help with any awkward numbers).

Questions to ponder	Resolution (rough approximates only)
How fast, in kilometres per hour, does human hair/human fingernails grow?	When I have my monthly haircut, on average, about 2 cm is trimmed off. Using 30 days per month and 24 hours per day, this works out at roughly $2 \div 30 \div 24 = 0.0028$ cm/h, or $0.0028 \div 100 \div 1000 = 0.000\,000\,028$ km/h.
Roughly how many cigarettes are smoked annually in the UK?	If roughly one adult in six in the UK smokes, that is about 8 million people. Assume that the average consumption of cigarettes among smokers is around 10 per day. Then the total number of cigarettes smoked annually is 8 million $\times$ 10 $\times$ 365 or roughly 30 billion.
Could you fit the entire population of the world on the Isle of Wight?	It will help to know that the approximate area of the Isle of Wight is 380 km^2 (or about 147 square miles). If we assume that everyone is standing up, you might fit four people into each square metre. There are 1 million square metres in one square kilometre, so the total number of people that could be accommodated is 380 $\times$ 1 000 000 $\times$ 4 or roughly 1.5 billion people. This is only just over 20% of the current population of the world, so it would seem to be theoretically impossible, although in practice I doubt if everyone would accept the invitation to try anyway!

Nugget: scientific notation

You may have spotted earlier the rather odd-looking number, 9.1094×10^{-31}. It may not have been obvious to you that this is an extremely tiny number. In fact this is a number so tiny it would be difficult to write out using conventional decimal notation. In fields such as science and astronomy, very large and very small numbers crop up often. In order to write them down clearly, an alternative notation has been devised called scientific notation (also known as standard form). Here is a simple example of scientific notation.

The number 2×10^3 means 2 with three zeros after it, or in other words 2000. The number 5×10^4 means 5 with four zeros after it, or in other words 50 000. In order to describe a tiny number less than 1, we use negative powers of 10. So 7×10^{-2} means 7 divided by 10^2 and is equivalent to 0.07. The number 6×10^{-5} means 6 divided by 10^5 and is equivalent to 0.000 06.

Here are two more estimations for you to think about.

▶ After a portion of clear plastic food-wrap has been used once, most people simply discard it. I have a friend who washes these little sheets of plastic after use, hangs them out to dry and then recycles them. This has never seemed to me to be a particularly economic use of her time! Assuming that this act of eco-enthusiasm takes my friend roughly 3 minutes per portion, and assuming also that a new 30-m roll of food-wrap costs about £2, estimate how much money she is saving per hour?

▶ In a newspaper article it was claimed that roughly 1 million Barbie dolls had been sold worldwide over the past 40 years and that this was at an average rate of roughly 2 every second. Are these two facts consistent with each other?

Nugget: coincidences

Recently, I flew on an aeroplane from Birmingham, UK, to the beautiful Canary island of Lanzarote. The seat beside me was occupied by a very pleasant stranger called Barry. When I told

Barry that I was interested in topics such as luck and coincidence, he pointed out that it was something of an amazing coincidence that he and I should have booked a flight from the same airport to the same destination, flying at exactly the same time and with our seats next to each other. But is this really a true coincidence? After all, the likelihood is that *someone* would have occupied that seat if it hadn't been Barry. A true coincidence would have been if I had already known Barry well and we had both booked the flights completely independently of each other.

RISK

Can you estimate the risk of being involved in a road accident, contracting a fatal disease, getting in the way of a terrorist's bullet...? Most people find it difficult to quantify risk and use such calculations to make sensible decisions. For example, it has been observed that in a week immediately following a serious air accident, some people prefer to travel by road on the grounds of safety. However, based on measurements of safety in terms of the number of deaths per passenger mile travelled, car travel is over 100 times more dangerous than air travel. Travelling by motorcycle, bicycle or on foot is yet more dangerous than travelling by car! So, using comparisons based on comparable distances travelled, air travel is actually much safer than road travel.

Perhaps one reason why risk is hard to comprehend is that it involves small probabilities arising from calculations involving big numbers (such as the total number of kilometres travelled by passengers on trains over the period of a year, and so on).

Other examples of dodgy deductions about risk are as follows:

▶ A few years ago, a man from Chicago was overheard to say: 'Jenny and I have cancelled our trip to Ireland, what with all the terrorists.' In fact, even at the height of the Troubles in Northern Ireland in the 1980s, the number of gun deaths per head of population in Chicago was roughly ten times that of Belfast!

▶ People sometimes believe that having two possible outcomes means that they are necessarily equally likely

('I will either pass my driving test or fail it, so I must have a 50–50 chance of passing'). This line of argument may be true of tossing a coin, where the two outcomes of 'heads' and 'tails' really are equally likely, but not necessarily true in other situations. For example, imagine that you are rolling a die once, hoping to get a 6. Can you see why the following argument is dodgy? 'I will either succeed in rolling a 6, or I won't. That's just two chances, so I should have a 50% chance of rolling a 6.'

▶ Risk comes up in weather forecasting where you might hear that there is a 40% chance of rain next Friday. What exactly does this mean? Let us suppose that it did turn out wet on Friday. People might claim that there was therefore a 100% chance of rain because it happened. However, prediction is about making claims into the future and these are always uncertain by their nature – it is easy to have 100% hindsight!

If you meet a stranger, what is the chance that you share a common acquaintance? This is a probability that most people tend to underestimate. The reason is that these two people are not just randomly selected from the 7 billion people inhabiting the world. For a start, the very act of meeting each other means that you are sharing the same tiny part of the globe. Also, the circumstances of the meeting will narrow the possibilities even further (it might be an event linked to work or some social event or travel arrangement, which suggest other characteristics that the two people might have in common). In general, the significance of coincidences is a greatly over-rated phenomenon that can be explained by the fact that many 'unlikely' events are actually more likely than you'd think. For example, suppose there are 25 strangers in a room. What is the probability that two or more of them share the same birthday? Most people would say that this was very unlikely and might come up with a probability estimate of less than 1 in 10 (presumably based on a calculation such as 25/365). However, it turns out that the likelihood of at least one shared birthday among 25 people is actually greater than 50–50! If you want to know why this is, think about the possible number of pairings that there are between any two people, any of which might yield a shared

birthday. You could start by numbering each person 1, 2, 3, ...
25 and then write down the possible pairings, as follows:

1/2, 1/3, ... 1/25 (there are 24 pairings here)
2/3, 2/4, ... 2/25 (there are 23 pairings here)
3/4, 3/5, ... 3/25 (there are 22 pairings here)
...
23/25, 24/25 (there are 2 pairings here)
24/25 (there is 1 pairing here)

By setting the problem out in this way, you can see that there
are actually many more than 24 possible pairings (there are
actually 24 + 23 + 22 + ... + 2 + 1 = 300 pairings), so the number
of possible ways of achieving success (i.e. of getting a shared
birthday) is greater than you might initially have thought.

Nowhere does an understanding of risk crop up more forcefully
than in the area of health and the treatment of illness. A major
problem here is that if you have been ill and then get better it is by
no means clear whether this happy outcome was due to the 'cure'
or whether you would just have got better anyway. There is an old
saying that 'properly treated, you can get rid of a cold in seven days
but left to its own devices it can drag on for a week'. A key problem
with processing data in the area of risk and coincidence is that
most people tend to talk about their successes but remain silent on
their failures. Added to this, there is usually enough random success
to justify almost anything to someone who wants to believe!

What sort of statistical questions can be asked?

There seem to be three basic types of question that crop up regularly
in statistics. These are listed below, with an example of each.

(A) CAN YOU SUMMARIZE THE DATA?

Looking at a lot of facts and figures does not always provide you
with a clear picture of what is going on. It is often a good idea to
find a way of summarizing the information – perhaps by reducing
the figures to just one representative figure. For example, back
in the 1950s it was not well understood how mothers' smoking

in pregnancy could adversely affect their child's birth weight and general health. As the message began to get across, a few women began to cut down or stop smoking altogether during their pregnancies. But in these early days, it was difficult to detect much difference in women's smoking behaviour. In order to get a sense of the extent to which women smokers were starting to reduce their smoking during pregnancy, 1000 women smokers might have been asked to record the number of cigarettes they smoked before and during pregnancy. The raw data would yield two sets of 1000 numbers each – too much information to take in at a glance. An obviously useful way of summarizing the data in this case would be to calculate the average number of cigarettes smoked before and during pregnancy.

There are a number of statistical techniques for summarizing data – for example, 'summary statistics' such as graphs and averages. You can read about these in Chapters 2, 3, 4 and 5.

(B) IS THERE A SIGNIFICANT DIFFERENCE BETWEEN THESE TWO SETS OF RESULTS?

Many of the decisions we make are based on our judgement of whether one thing performs better, lasts longer or offers better value for money than another. It is easy enough to show that one set of results is different from another, but just how significant that difference might be is less easy to decide on. So, following on from the previous example, let's just make up some numbers here and say the researchers found that:

Average daily number of cigarettes smoked

before pregnancy	*.....15.2*
during pregnancy	*.....14.9*

Clearly there is a difference between these two results (although it would be very surprising if they had turned out to be identical) but it is by no means obvious whether we can place any great significance on the degree of difference.

This is where the idea of 'significance tests' comes in. These give a way of assessing the differences that you observe to help you decide whether or not two results are 'significantly different' from each other. This idea is discussed in Chapter 14.

Nugget: an aside on smoking in pregnancy

These days (in the Western developed world in the 21st century), most mothers choose not to smoke during their pregnancy. So, in practice, the differences in smoking patterns would be much greater than those suggested in the made-up figures above.

(C) IS THERE A CLOSE RELATIONSHIP BETWEEN THE TWO THINGS UNDER CONSIDERATION?

Sometimes we are less interested in differences between two sets of results than in exploring the possible relationship between two things. For example, the researchers on smoking may wish to try to identify the main factor, or factors, that are linked to smoking. They may use their data to test whether there was a close connection between, say, a person's level of *psychological stress* and the number of cigarettes smoked, or if *disposable income* provided a stronger link with smoking. (Of course, the factors of stress and income are themselves closely intertwined and part of the task for the statistician is to try to isolate each of the factors that they choose to consider.)

This idea of a relationship between two things is discussed in Chapters 11 and 12. Chapter 11 is concerned with describing the sort of relationship, while Chapter 12 deals with measuring how strong the relationship is.

How are statistical questions investigated?

Data are usually collected in response to a question. The decision as to which technique is to be used will depend on the sort of question that has been asked at the start of the investigation. Posing a clear question is normally the first stage of any statistical work. For example, a doctor might be interested in investigating whether a new treatment for asthmatics is more effective than the old one. A sales manager may wish to find out which of a variety of different company advertisements was actually the most successful. A teacher could plan to catalogue and summarize the most common errors committed by students in a particular topic. These are the sorts of questions which give a purpose and a direction to statistical work and they breathe life and interest into the subject.

Nugget: data

The word 'data' refers to facts and figures or indeed to any information that people might use to inform their judgements. The singular of data is 'datum', which means a single fact or figure. So, data is a plural word and that is the way it has been used here. However, in many textbooks it is used in the singular; for example, 'Where is the data?'

As was indicated in the Introduction, there are four clearly identifiable stages in most statistical investigations, which can be summarized as follows.

THE STAGES OF A STATISTICAL INVESTIGATION

▶ Stage 1: pose a question

▶ Stage 2: collect relevant data

▶ Stage 3: analyse the data

▶ Stage 4: interpret the results

Not surprisingly, there are different statistical techniques suitable for the different stages. For example, issues to do with choosing samples and designing questionnaires will tend to crop up at the second stage, 'collect relevant data'. The third stage of analysing the data will probably involve various calculations (perhaps finding the average or working out a percentage) while the final stage of interpreting the results will raise questions such as whether the relationship under consideration is likely to be one of cause and effect. Table 1.1 summarizes some of these connections.

Table 1.1 Connecting the stages and the tools of statistics

Stages of an investigation	Tools and techniques of statistics
1 Pose a question	
2 Collect relevant data	▸ Choose a sample
	▸ Design a questionnaire
	▸ Conduct an experiment
3 Analyse the data	▸ Calculate a percentage
	▸ Calculate a mean
	▸ Draw a helpful graph
4 Interpret the results	▸ Make a prediction
	▸ Make a comparison
	▸ Test for cause and effect

Although this book has been written with these stages clearly in mind, the chapters do not follow rigidly through them in sequence. For a start, there are no chapters specifically concerned with Stage 1, 'pose a question'. Also, it seemed sensible to devote the first part of the book (Chapters 1 to 6) to providing a basic explanation of some of the simpler statistical graphs and calculations. For example, Chapter 2 explains some basic ideas of maths that you will draw on later in the book. Thereafter, the second stage of 'collecting the data' is dealt with in Chapters 7 and 8 (on sampling and sources of data). Stage 3, 'analysing the data', deals with the tools and techniques of the subject and is central to Chapters 3 and 4 (on graphs), Chapter 5 (summary statistics), Chapter 9 (spreadsheets) and Chapter 10 (reading tables of data). Lastly, the interpretation phase of statistical work is discussed in Chapter 6 (Lies and statistics) and in the final four chapters of the book (covering regression, correlation, probability and statistical tests of significance). This is summarized in Table 1.2.

Table 1.2 Themes and chapters of the book

Theme	Chapter	
General introduction	1	Introducing statistics
	2	Some basic maths
Pose a question		
Collect relevant data	7	Choosing a sample
	8	Collecting information
Analyse the data	3	Graphing data
	4	Choosing a suitable graph
	5	Summarizing data
	9	Spreadsheets to the rescue
	10	Reading tables of data

Theme	Chapter
Interpret the results	

For a more comprehensive guide to the key ideas of statistics and how they are connected, turn now to the Appendix at the back of the book, entitled 'Choosing the right statistical technique'. Here you will find a flow diagram which takes you, in a sensible sequence, through the sort of questions that users of statistics tend to ask. The diagram indicates the common statistical techniques that are required at each stage and where they can be found in this book. If at any time, when reading later chapters of the book, you are uncertain where the ideas fit into a wider picture, you may find it helpful to turn to this diagram again.

Choosing and using your calculator

Finally, you may have been wondering whether it would be a good idea to use a calculator (or perhaps buy a new one) to support your learning of statistics. Take courage! The calculator has a very positive role to play. In the past, a feature of studying statistics was the time students had to spend performing seemingly endless calculations. Fortunately, those days have gone, for, with the support of even a basic four-function calculator, the burden of hours spent on arithmetic is no longer necessary. These simple calculators are both cheap and reliable. However, if you are prepared to spend just a little more and buy a rather more sophisticated calculator, with customized statistical keys, the proportion of your time spent actually focusing on the central ideas of statistics will greatly increase. You may be thinking of buying a calculator, in which case the keys that you will find particularly useful in statistical work are shown below. Don't worry if they seem unfamiliar, as these topics will be explained in later chapters.

It is possible that some extra statistical keys are available on your calculator, but the chances are that these are just variants of the ones listed below.

$\bar{x}$	(x bar) calculates the average, i.e. the mean (see Chapters 2 and 5)
Σx	(sigma x) adds all the x values together (see Chapter 2)
Σx^2	(sigma x squared) adds the squares of the x values
n	finds the number of values already entered
$x\sigma_n$	(x sigma n), sometimes written as σ_n, finds the standard deviation of the x values (see Chapters 2 and 5)
r	finds the coefficient of correlation (see Chapter 12)
a, b	finds the linear regression coefficients (see Chapter 11)

(See page 30 for discussion of the two symbols for sigma.)

Although this book has been written with such a calculator in mind, explanations will also be included based on pencil and paper methods only.

Finally, you probably have access to a computer with a spreadsheet package. If so, this will be a most valuable tool for analysing data and gaining a better grasp of statistical ideas. Don't worry if you have never used a spreadsheet before – in Chapter 9 you will be provided with a basic introduction, from scratch, of how to use one.

The next chapter deals with a few important topics in maths which should help you to make the most of some of the later chapters. These are simple algebra, graphs and some of the mathematical notations commonly found in statistics textbooks.

To find out more via a series of videos, please download our free app, Teach Yourself Library, *from the App Store or Google Play.*

2

Some basic maths

In this chapter you will learn:

▶ *the basics of algebra, including the meaning of symbols and some useful definitions*

▶ *how to draw and interpret graphs*

▶ *some important statistical notation, including the symbols used for 'sum of' and 'mean'.*

If you flick through the rest of this book, you will find that complex mathematical formulas and notations have been kept to a minimum. However, it is hard to avoid such things entirely in a book on statistics, and if your maths is a bit rusty you should find it worthwhile spending some time working through this chapter. The topics covered are algebra, simple graphs and statistical notation and they offer a helpful mathematical foundation, particularly for Chapters 5, 11 and 12.

Algebra

Algebraic symbols and notation crop up quite a lot in statistics. For example, statistical formulas will usually be expressed in symbols. Also, in Chapter 11 when you come to find the 'best-fit' line through a set of points on a graph, it is usual to express this line algebraically, as an equation.

Clearly it would be neither possible nor sensible to attempt to cover all the main principles and conventions of algebra in this short section. Instead, your attention is drawn to two important aspects of algebra which are sometimes overlooked in textbooks. The first of these is very basic – *why do we use letters at all* in algebra and what do they represent? The second issue concerns *algebraic logic* – the agreed conventions about the order in which we should perform the operations of +, –, × and ÷.

WHY LETTERS?

Can you remember how to convert temperatures from degrees Celsius (formerly known as centigrade) to degrees Fahrenheit? If you were able to remember, the chances are that your explanation would go something like this:

> *Suppose the original temperature was, say, 20°C. First you multiply this by 1.8, giving 36, and then you add 32, so the answer is 68°F.*

The explanation above, which is perfectly valid, is based on the common principle of using a *particular* number, in this case a temperature of 20°C, to illustrate the *general* method of doing the calculation. It isn't difficult to alter the instructions in order to do the calculation for a different number, say 25 – simply replace the 20 with 25 and continue as before. What we really have here is a formula, in words, connecting Fahrenheit and Celsius. A neater way of expressing it would be to use letters instead of words, as follows:

$$F = 1.8C + 32$$

(where F = temperature in degrees Fahrenheit and C = temperature in degrees Celsius).

Reading the formula aloud from left to right, you would say 'F is equal to one point eight C plus thirty-two'.

The formula is equivalent to the word description shown earlier but here the letter C stands for whatever number you wish to convert from °C to °F. The F is the answer you get expressed in °F. The choice of letters for a formula is quite arbitrary – X and Y are particular favourites – but it makes sense to choose letters that make it is easy to remember what they stand for (hence F for Fahrenheit and C for Celsius in the formula shown above). The main point to be made here is that, in algebra, each letter is a sort of place holder for whatever number you may wish to replace it with. A formula such as the one above provides a neat summary of the relationship that you are interested in and shows the main features at a glance.

Nugget: fruit salad algebra

The term 'fruit salad algebra' refers to a common misconception about the meaning of letters in algebra. Teachers sometimes introduce apples and bananas to represent the letters a and b. Their aim is to help students distinguish between letters in algebra and not mix them up. However, drawing on this sort of lesson, when students see an algebraic term such as $5a$, they sometimes mistakenly think this refers to five apples. In fact, the letter a refers to the unknown number in the question and $5a$ means five times that number. So, $5a$ might mean five times an unknown number of apples rather than simply five apples. This misconception, for obvious reasons, is known as 'fruit salad algebra'.

There are certain conventional rules in algebra which you need to be clear about. Firstly, have a look at the following examples where a number and a letter are written close together:

$$5X; 1.8C; -3y$$

Although it doesn't actually say so, in each case the two things, the number and the letter, are to be *multiplied* together. Thus, $5X$ really means '5 times X'. Similarly, $1.8C$ means '1.8 times C', and so on.

Let us now go back to the temperature conversion formula. The table below breaks it down into its basic stages, with an explanation added at each stage.

Table 2.1 The temperature formula, $F = 1.8C + 32$, stage by stage

Meaning	
$F =$	The temperature, in degrees F, is equal to...
$1.8C$	... 1.8 times the temperature in degrees C...
$+32$	... plus 32

Exercise 2.1 Using the temperature formula

Use the formula $F = 1.8C + 32$ to convert the following temperatures into degrees Farenheit.

a 0°C

b 100°C

c 30°C

d -17.8°C

[Comments at the end of the chapter]

A formula is really another word for an *equation*. Usually a formula is written so that it has a single letter on the left of the equals sign (in this case the F) and an expression containing various numbers and letters on the right. We say that the equation $F = 1.8C + 32$ is a *formula for F* (F is the letter on the left of the equals sign) *in terms of C* (C is the only letter, in this case, on the right-hand side of the equals sign).

To *satisfy* a formula or an equation is to find values to replace the letters which will make the two sides of the equation equal. For example, the values $C = 0$, $F = 32$ will satisfy the equation above. We can check this by *substituting* the two values into the equation, thus:

$$F = 1.8C + 32$$
$$32 = 1.8 \times 0 + 32 \ldots \text{which is true!}$$

If you feel you need a little practice at this idea, try Exercise 2.2 now.

Exercise 2.2 Satisfying equations

Which of these pairs of values satisfy the equation $Y = 3X - 5$?

a $X = 3$, $Y = 4$

b $X = 0$, $Y = 5$

c $X = -2$, $Y = -11$

[Comments at the end of the chapter]

Nugget: equation and expression

Students often mix up the meanings of the words *equation* and *expression*. The key to distinguishing between them is to observe that the word equation contains, in its first four letters, a reference to 'equals'. So, an equation must contain an equals sign, whereas an expression will not. Another key difference between these two algebraic ideas is what you do with them. Equations exist to be solved (that is finding the value of the unknown letter) whereas all you can do with an expression is to simplify it.

ALGEBRAIC LOGIC

A second important convention in algebra is the *sequence* in which you are expected to do calculations within a formula.

In the previous example, notice that the temperature in °C is multiplied by 1.8 *before* you add the 32. In general, where you appear to have a choice of whether to add, subtract, multiply or divide first, the multiplication and division take precedence over addition and subtraction. This would be true even if the formula had been written the other way round, like this:

$$F = 32 + 1.8C$$

This may seem odd but the reason for the convention is simply that the formula would otherwise be ambiguous. So it doesn't matter which way round the formula is written; provided the multiplication is done first, the result is still the same. All scientific calculators are programmed to obey this convention of algebraic precedence, but generally speaking simple four-function calculators are not. If you have a calculator to hand, this would be a good opportunity to check out its logic, if you haven't already done so.

Exercise 2.3 will guide you through the necessary steps.

Exercise 2.3 Guess and press

Guess what result you think you would get for each of the following key sequences and then use your calculator to check each guess.

KEY SEQUENCE	YOUR GUESS	PRESS AND CHECK
2 ⊠ 3 ⊟ 4 ⊟		
2 ⊞ 3 ⊠ 4 ⊟		
2 ⊟ 3 ⊠ 4 ⊟		
2 ⊟ 3 ⊞ 4 ⊟		

[Comments at the end of the chapter]

As a result of the previous exercise, you should now be clear about the nature of the 'logic' on which your calculator has been designed. However, whether your calculator is programmed with 'arithmetic' logic or 'algebraic' logic, written algebra obeys the rules of algebraic logic. We can take as an example the second key sequence shown in Exercise 2.3. If you were to read this sequence to a human calculator, the chances are that they would produce the answer 20 ($2 + 3 = 5$, times $4 = 20$). However, suppose the question is now asked in the following slightly different way:

Find the value of $2 + 3X$, when $X = 4$.

This still produces the same calculation (2 + 3 × 4), but the rules of algebra require that the sequence of the operations gives higher priority to the multiplication than the addition. If the spacing is changed slightly, this idea is easier to grasp, thus:

$$2 + 3 \times 4$$

When the multiplication is done first, the algebraically 'correct' answer is 14.

There is not room here to develop this notion of the logic of algebraic precedence further (i.e. the priorities given to each operation), but you will find that a good understanding of the ideas described above is helpful if you wish to handle algebraic expressions and equations successfully.

You might be wondering what happens if you want a formula involving multiplication and addition but you wish the addition to be done before the multiplication. This is where the use of brackets comes into algebra. An example of this is where the temperature conversion formula is rearranged so that it will convert from degrees Fahrenheit into degrees Celsius. Remember that, if you reverse a formula, not only do you have to reverse the operations (× becomes ÷ and + becomes –) but you must also reverse the sequence of the operations. Exercise 2.4 will give you a chance to work this out for yourself.

Exercise 2.4 Reversing a formula

The middle column of Table 2.2 gives the two stages in the formula for converting temperatures from °C to °F.

a Fill in the blank third column to show the two stages (in the correct sequence) needed to convert temperatures from °F to °C.

Table 2.2 Converting to °C

Sequence	Converting from °C to °F	Converting from °F to °C
First	multiply by 1.8	
Second	add 32	

b Use your answer to part (a) to write down a formula which would allow you to convert temperatures from °F to °C.

[Comments at the end of the chapter and continued below]

There are certain problems if we write the new conversion formula as:

$$C = F - 32 \div 1.8$$

If this formula were to be entered directly into a calculator with algebraic logic, it would give the 'wrong' answer. The reason for this is that, because division has a higher level of precedence than subtraction, a calculator with algebraic logic would divide 32 by 1.8 before doing the subtraction. Putting $F - 32$ in brackets, as shown below, takes away the ambiguity and forces the calculator to perform the calculation in the required sequence.

$$C = (F - 32) \div 1.8$$

To end this section, here are a few of the common algebraic definitions that you may need to be familiar with.

▶ **Expression:** *$3X^2 - 4X + 5XY + 3X$ is an example of an expression.*

▶ **Term:** *There are four terms in the previous expression. They are, respectively, $3X^2$, $-4X$, $5XY$ and $3X$.*

▶ **Sign:** *The sign of a term is whether it is positive or negative. Only the second of these four terms is negative; the others are positive. When we are writing down a positive term on its own we don't normally bother to write the '+' sign before it. Thus we would write $5XY$, rather than $+5XY$.*

▶ **Term type:** *This refers only to the part of the term that is written in letters. Thus, the first term in the expression above is an 'X-squared' term, the second is an 'X' term and so on.*

▶ **Coefficients:** *The coefficient of a term is the number at the front of it. Thus, the coefficient of the first X term is -4. The coefficient of the XY term is 5, and so on. The coefficient tells you how many of each term type there are.*

▶ **Like term:** *Two terms are said to be 'like terms' when they are of the same term type. Thus, since the second and fourth terms in the expression above are both 'X terms', they are therefore 'like terms'. The phrase 'collecting like terms' describes the process of putting like terms together into a* single term. For example the second and fourth terms, -4X and 3X, can be put together simply as -X. This result arises from adding the two coefficients, -4 and 3, giving a combined coefficient of -1. We don't normally write this as -1X, but rather as -X.

▶ **Simplifying expressions:** *This is the general word for collecting like terms. It is a simplification in that the total number of terms is reduced to just one of each term type.*

▶ **Expanding:** *This normally refers to expanding out (i.e. multiplying out) brackets. Thus, the expression $5(2X + 3) - 3(6 - X)$ could be expanded out to give $10X + 15 - 18 + 3X$. The purpose of doing this is usually to enable further simplification. Thus, in this case we can collect together the two X terms and the two number terms to give the simplified answer $13X - 3$.*

▶ **Equations:** *Equations look rather like expressions except that they include an equals sign, '='. In fact, an equation can be defined as two expressions with an equals sign between them. The expression to the left of the equals sign is, for obvious reasons, called the 'left-hand side', or 'LHS', while the other expression to the right of the equals sign is the 'right-hand side' or 'RHS'. Here is an example of an equation: $2X - 3 = 4 - 6X$. Often, however, equations are written so that the RHS expression is zero. For example, $3 - 4X + X^2 = 0$.*

▶ **Solving equations:** *Equations usually exist in mathematics textbooks to be 'solved'. Solving an equation means finding the value or values for X (or whatever the unknown letter happens to be) for which the equation is true. For example, the simple equation $3X - 5 = 7$ holds true for one value of X. The solution to this equation is $X = 4$. This can be checked by 'substituting' the value $X = 4$ back into the original equation to confirm that the LHS equals the RHS. Thus, we get LHS $= 3 \times 4 - 5$, which equals 7. This is indeed the value of the RHS. If you were to try any other value for X say $X = 2$, the equation would not be 'satisfied'. Thus, LHS $= 3 \times 2 - 5$, which equals 1 not 7. The equation $3 - 4X + X^2 = 0$ holds true for two values of X, $X = 1$ and $X = 3$, so there are two solutions to this equation.*

And that is all the algebra we will touch on here. The next section deals with graphs and how equations can be represented graphically.

Graphs

Perhaps more than any other mathematical topic, the basic idea of a graph keeps cropping up in statistics. The reason for this is to do with how the human brain processes information. For most people, scanning a collection of figures, or looking at a page of algebra, doesn't reveal very much in the way of patterns. However, our brains seem to be much better equipped at seeing patterns when the information has been re-presented as a picture. The crucial difference is that in a picture (a graph or diagram, for example) each fact is recorded onto the page by its *position*, rather than by writing some peculiar, arbitrary squiggle (a number or a letter). In order to understand how graphs work, it is worth spending a few minutes thinking about the idea of position and what information you need to provide in order to define position exactly. Now do Exercise 2.5.

Exercise 2.5 ✕ marks the spot

Have a look at the cross, ✕, in the line above. How would you explain to someone else exactly where the ✕ is located on this page?

[Comments follow]

Basically there are three pieces of information that you need to provide in order to fix exactly where the ✕ is located. These are:

▶ which (original) point on the page you started measuring from – known as the 'origin';

▶ how far from the origin the point is across the page – the *horizontal* distance; and

▶ how far from the origin the point is *up* the page – the *vertical* distance.

In mathematics, graphs are organized around these three principles. The 'page' is laid out in a special way so that the starting point (the origin) is clearly identified – normally in the bottom left-hand corner. The two directions, horizontal and vertical, are normally drawn out as two straight lines and are known, respectively, as the X axis and the Y axis. Finally, in order to keep things as simple as possible, the two axes are provided with a scale of measure.

Nugget: order matters

For all those football fans out there, have a look at this result:

Manchester United 3, Arsenal 2

This result tells you how many goals each side scored, but it also tells you something else. You are able to deduce that the match was played in Manchester, and not in London. I know this because of the order in which the teams are listed. There is a convention that the first mentioned team is the home team. A similar sort of convention exists with the way that coordinates are written. For example, look at the coordinates (2, 3). The two numbers, 2 and 3, show their distance from the origin, but their order tells you which is which. The first number, 2, is measured in the horizontal direction and the second, 3, in the vertical direction. So to move from the origin to the point with coordinates (2, 3), take two steps to the right followed by three steps up.

So, provided the page is laid out with the origin clearly identified, and the axes are suitably scaled, the point marked with an × in Figure 2.1 can be described by giving just two numbers, in this case 10 and 20. These two numbers are called the coordinates of the point and they are usually written inside brackets, separated by a comma, like this:

$$(10, 20)$$

The first number inside the brackets, known as the X coordinate, measures how far you have to go along the X direction from the origin. The second number, the Y coordinate, gives the distance you go up the Y direction. Exercise 2.6 will give you practice at handling coordinates.

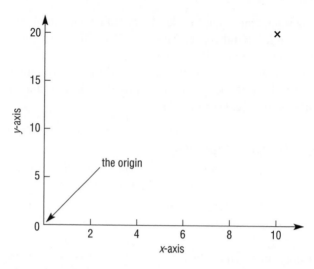

Figure 2.1 The axes of a graph

Exercise 2.6 Plotting coordinates

Plot the following coordinates on a graph.

(0, 10) (2, 12) (4, 14) (6, 16) (8, 18) (10, 20)

If you plot them correctly, the points should all lie on a straight line.

[Comments at the end of the chapter]

Having completed Exercise 2.6, it is clear just by looking at the pattern of points on the graph that there must be some special property in these coordinates which they all share. Have a look now at the coordinates and see if you can find any pattern in the numbers.

You may have spotted that, in each case, the Y coordinates are 10 more than their corresponding X coordinates. We can say this more neatly, using the language of algebra.

For each point, $Y = X + 10$

In fact, if a straight line were drawn through the points, this property ($Y = X + 10$) would apply to each and every one of the many possible points that you might wish to choose that lie on the line. Each of the points that you plotted in Exercise 2.6 is said to 'satisfy' the equation. For example, the point (4, 14) consists of an X value ($X = 4$) and a Y value ($Y = 14$).

Substituting (4, 14) into the equation $Y = X + 10$

we get $14 = 4 + 10$

To satisfy the equation means that this pair of values has produced a true statement. The full significance of this result is summarized below.

Graph	Algebra
If a point lies on a line...	... the coordinates of the point 'satisfy' the equation of the line

This is the crucial connection between graphs and algebra.

In the last example, you started with a set of points and deduced the equation of the line connecting them. In the next exercise, you will be asked to do this in reverse – i.e. you will be given a new equation, $Y = 2X - 3$, and asked to draw its graph by plotting various points.

Exercise 2.7 Drawing the graphs of equations

a Check that the point (2, 1) 'satisfies' the equation $Y = 2X - 3$.

b Complete the following coordinates so that they satisfy the same equation: (0, ?) (1, ?) (?, 3) (4, ?).

c Plot the point (2, 1) and the four points that you have worked out in (b) onto a graph. They should all lie on a straight line whose equation is $Y = 2X - 3$.

[Comments at the end of the chapter]

You will have already noticed that, as with the previous example, the pattern of points appears as a straight line. It is worth commenting here that equations such as $Y = X + 10$ and

$Y = 2X - 3$ are known as *'linear'* equations for this reason – that is, when they are plotted onto a graph, they produce a *straight line*.

Finally, have a look at where the $Y = 2X - 3$ line passes through the Y axis. It is significant that it intersects the Y axis at the same value as the number term (-3) in the equation. This is a useful fact to remember about linear equations. The value -3 is known as the 'Y intercept'. The other number in the equation is the one associated with the X, in this case the number 2, and it will tell you how steep the line is – known as the 'slope' or 'gradient' of the line. It is a measure of how far the line goes up in the Y direction for every unit distance it goes across in the X direction. Expressed in more mathematical language, we can summarize all this by the following statement:

> The general linear equation, $Y = a + bX$, has a *slope* of b and an *intercept* of a.

You may find it helpful to see a practical example of these ideas, so here is how it applies to working out the bill that you might expect to pay your plumber who has a callout fee of £50 and on top of that charges at a rate of £40 per hour.

The explanation is shown below in three different forms: first in words, then in a formula and finally in the form of a graph.

HOW TO CALCULATE THE PLUMBER'S BILL

▶ (a) in words

You pay a callout fee of £50 plus £40 per hour.

▶ (b) in a formula

$B = 50 + 40H$ (where B is the total bill, in pounds, and H is the number of hours).

▶ **(c) in a graph**

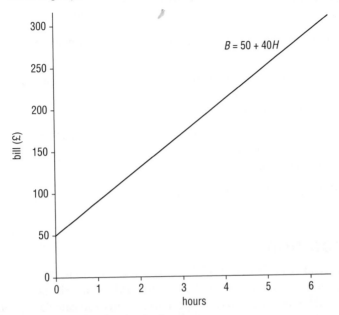

Figure 2.2 Graphing the plumber's rates

This graph has produced the familiar 'linear' pattern. As you can see, the intercept of the graph gives the fixed part of the formula (the £50), whereas the slope is a measure of the charge rate per hour. Provided you know how many hours you have billed for, you can use the graph to read off the bill. The next exercise asks you to do this now.

Exercise 2.8 Work out your bill from the graph

Suppose the plumber charged you for five hours of work. From the graph, read off what bill you would expect to have to pay.

[Comments at the end of the chapter]

Now to end this section on graphs, here are a few questions to get you thinking a bit more deeply about their properties.

Exercise 2.9 Some further questions about graphs

a If $Y = a + bX$ is the general form of the linear equation, what does the general equation $Y = mX + c$ represent?

b A series of linear equations all have the same intercepts but different slopes. Draw a sketch of what you think they would look like if they were plotted on the same axes.

c A series of linear equations all have the same slopes but different intercepts. Draw a sketch of what you think they would look like if they were plotted on the same axes.

[Comments at the end of the chapter]

Notation

There are one or two special symbols and notations in statistics that are in common use and are well worth familiarizing yourself with. If you have a specialized statistical calculator, you will already have run into some of these. The two described here are Σ and $\overline{X}$.

(CAPITAL) SIGMA Σ

The first important symbol is the Greek letter capital sigma, written as Σ. In order to keep things interesting for you, statisticians have not one but two symbols called 'sigma' in regular use. The lower case Greek letter sigma is written as 'σ' and it is normally used in statistics to refer to a measure of spread, called the 'standard deviation' (which is explained in Chapter 5).

This symbol, Σ, is an instruction to add a set of numbers together. So, ΣX means 'add together all the X values'. Similarly, ΣXY means 'add together all the XY products'.

Table 2.3 shows all the X and Y values which you plotted as coordinates in Exercise 2.7.

Table 2.3 Summing the X and Y values separately

X	Y
0	-3
1	-1
2	1
3	3
4	5
$\Sigma X = 0 + 1 + 2 + 3 + 4 = 10$	$\Sigma Y = (-3) + (-1) + 1 + 3 + 5 = 5$

To find ΣXY, it is necessary to calculate all the five separate products of X times Y and then add them together, thus:

Table 2.4 Summing the XY products

X	Y	XY
0	-3	0
1	-1	-1
2	1	2
3	3	9
4	5	20
		$\Sigma XY = 0 + (-1) + 2 + 9 + 20 = 30$

Exercise 2.10 Calculating with Σ

Use the data from the example above to calculate the following:

a ΣX^2

b ΣY^2

c $\Sigma (X + Y)$

[Comments at the end of the chapter]

X BAR, $\bar{X}$

The symbol $\bar{X}$ is pronounced 'X bar' and refers to the mean of the X values. A 'mean' is the most common form of average, where you add up all the values and divide by the number of values you had. To take the previous example, the five X values are 0, 1, 2, 3, and 4. In this example, the mean value of X is written as:

$$\bar{X} = \frac{\Sigma X}{n}$$

Thus, $\overline{X} = \dfrac{0 + 1 + 2 + 3 + 4}{5} = \dfrac{10}{5} = 2$

Comments on exercises

▶ Exercise 2.1

a 32°F (freezing point of water)

b 212°F (boiling point of water)

c 86°F

d -0.04°F (or roughly zero)

▶ Exercise 2.2

We can substitute each of these pairs of values in turn into the equation to see if they give a correct result.

$$Y = 3X - 5$$

a $X = 3$, $Y = 4$	$4 = 3 \times 3 - 5 \dots$	which is true
b $X = 0$, $Y = 5$	$5 = 3 \times 0 - 5 \dots$	which is not true
c $X = -2$, $Y = -11$	$-11 = 3 \times (-2) - 5 \dots$	which is true

▶ Exercise 2.3

In terms of their inner 'logic', calculators come in two basic types. At the cheaper end of the market are the so-called 'arithmetic' calculators which perform calculations from left to right in the order in which they are fed in. The more sophisticated (and, generally, more expensive) 'algebraic' calculators are programmed to obey the conventions of algebraic precedence described earlier. Depending on which calculator you have, here are the results you are likely to get.

Key sequence	'Arithmetic' answer	'Algebraic' answer
2 ⊠ 3 ⊟ 4 ⊟	2	2
2 ⊞ 3 ⊠ 4 ⊟	20	14
2 ⊟ 3 ⊠ 4 ⊟	-4	-10
2 ⊟ 3 ⊡ 4 ⊟	-0.25	1.25

Nugget: pronouncing 'arithmetic'

There are two pronunciations of the word 'arithmetic'. When the emphasis is on the second syllable (the '-ith') the word refers to calculations with numbers. But when the emphasis is on the third syllable (the '-met') the word has a different meaning. Here it refers to a type of logic built into the cheaper models of calculator where calculations are performed from left to right.

▶ **Exercise 2.4**

a

Sequence	Converting to °F	Converting to °C
First	multiply by 1.8	subtract 32
Second	add 32	divide by 1.8

b A suitable formula would be: $C = (F - 32) \div 1.8$, which may also be written as:

$$C = \frac{F - 32}{1.8}$$

▶ **Exercise 2.6**

The six points are located on the graph as shown below.

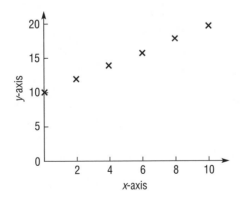

▶ **Exercise 2.7**

a Substituting the values $X = 2$ and $Y = 1$ into the equation $Y = 2X - 3$, we get:

$1 = 2 \times 2 - 3$ ... which is true.

b The other four coordinates are $(0, -3)$, $(1, -1)$, $(3, 3)$ and $(4, 5)$.

c The five coordinates and graph of the equation should look like this:

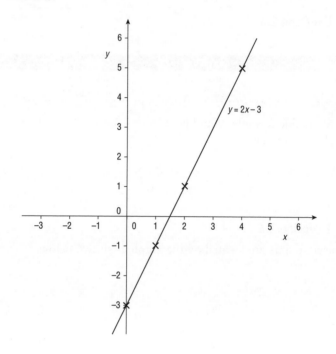

▶ Exercise 2.8

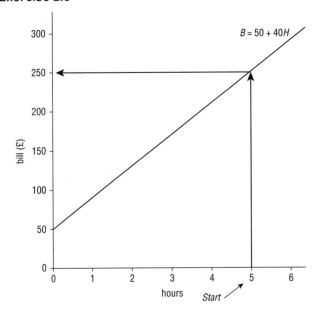

$B = 50 + 40H$

bill (£)

hours   *Start*

To make an estimate of your bill, follow the procedure shown on the graph above. The stages involved are described below.

a Start at 5 hours on the horizontal axis.

b Draw a line vertically upwards until it meets the graph.

c From this point on the graph, draw a line horizontally across to the vertical axis.

d Read off this axis the bill you will have to pay (£250).

▶ Exercise 2.9

a These two equations may look different but they actually have the same form. In the second version (which is popular in some maths textbooks) the 'm' is the letter used to represent the slope and the 'c' is used for the intercept. The ordering of the two terms to the right-hand side of the equals sign is quite arbitrary and of no mathematical significance.

b These lines all have the same intercept but different slopes.

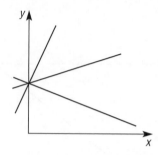

c These lines have all got the same slope but different intercepts.

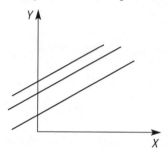

▶ **Exercise 2.10**

The table below sets out the calculations that are required.

X	Y	X²	Y²	X + Y
0	-3	0	9	-3
1	-1	1	1	0
2	1	4	1	3
3	3	9	9	6
4	5	16	25	9
		$\sum X^2 = 30$	$\sum Y^2 = 45$	$\sum(X + Y) = 15$

Key ideas

This chapter introduced three basic mathematical ideas which underpin a lot of statistical work.

▶ Algebra: *The section on algebra looked at two important aspects of this branch of mathematics – why we bother to use letters at all and the idea of algebraic logic (which controls the sequence in which algebraic operations are performed).*

▶ Graphs: *The section on graphs explained how coordinates are plotted and introduced the terms slope and intercept of linear (i.e. straight line) equations. An important link was made between an algebraic equation and what it looks like when drawn as a graph.*

▶ Notations: *The final section explained two useful notations in statistics – the capital sigma (Σ), which means that a set of numbers are to be added together, and the use of a bar over a letter (for example, $\overline{X}$) to signify the mean of a set of values.*

To find out more via a series of videos, please download our free app, Teach Yourself Library, *from the App Store or Google Play.*

3

Graphing data

In this chapter you will learn:

▶ *how to draw and interpret graphs and diagrams, including bar charts, pie charts, pictograms, histograms, stemplots, scattergraphs and time graphs.*

This chapter introduces the main features of the following graphs: bar charts, pie charts, pictograms, histograms, stemplots, scattergraphs and time graphs. Boxplots are explained in Chapter 5. Related issues to do with *when* these graphs might be usefully drawn and *how they might be interpreted* are discussed in Chapter 4, so try not to get too interested at this stage in where the data came from and the patterns they may suggest to you!

Nugget: letting the machine decide

Increasingly these days, graphs, charts and plots are created on a computer rather than by using pencil and paper. There is no doubt that, by using a computer, you can achieve a most attractive chart quickly and easily. However, there is a downside. The really useful contribution offered by the machine is the technical one of transforming the data into the chart of your choice. What it cannot do is decide whether the particular data set is appropriate for presenting visually, or whether one particular chart is a helpful way of representing this information. My advice is to make good use of the time you save by using a computer to make sure that the chart is both sensible and appropriate and that the data are correctly collected and entered into the machine.

Bar charts

The bar chart is probably the most familiar of all the graphs that you are likely to see in newspapers or magazines. Figure 3.1 is a typical example, which summarizes the areas in square kilometres (km^2) of the seven continents of the world.

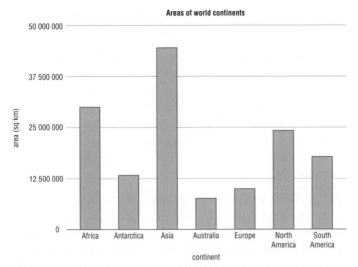

Figure 3.1

Table 3.1 shows the data from which the bar chart has been drawn.

Table 3.1 World continents by area

Continent	Area in square kilometres
Africa	30 370 000
Antarctica	13 720 000
Asia	43 820 000
Australia	9 008 500
Europe	10 180 000
North America	24 490 000
South America	17 840 000

Source: https://en.wikipedia.org/wiki/Continent#Area_and_population

The chief virtue of a bar chart is that it can provide you with an instant picture of the relative sizes of the different categories. This enables you to see some of the main features from a glance at the bar chart – for example in this case you can see which are the largest and smallest land masses. There are a number of general features of a bar chart that are worth drawing attention to.

First, you can see that the bars have all been drawn with the *same width*. This makes for a fair comparison between the bars, in that the height of a particular bar is then a fair measure of the value corresponding to that category.

Next, notice that it has been drawn with *gaps* between the adjacent bars. The reason for this is to emphasize the fact that the different continents listed along the horizontal axis are *separate categories*. This contrasts with a seemingly similar type of graph called a histogram, where the horizontal axis is marked with a continuous number scale and adjacent bars are made to touch. We will return to this point later in the chapter.

Now, have a look at the *order* in which the various continents have been placed along the horizontal axis of the bar chart. You may have wondered why this particular order has been chosen but in this case it is simply alphabetical order of the names of the continents. A more useful ordering would be to arrange the data by size of continent – in other words, the tallest bar to the left and the shortest to the right.

Another tweak worth considering would be to simplify the numbers on the vertical axis by expressing the areas in millions. This means, for example, that the area of the land mass of Africa would be expressed as 30.4 million km^2, rather than as 30 370 000 km^2. Both of these last two adjustments have been made in Table 3.2 and Figure 3.2.

Table 3.2 World continents by area, re-presented

Continent	Area (sq km) million
Asia	43.8
Africa	30.4
North America	24.5
South America	17.8
Antarctica	13.7
Europe	10.2
Australia	9.0

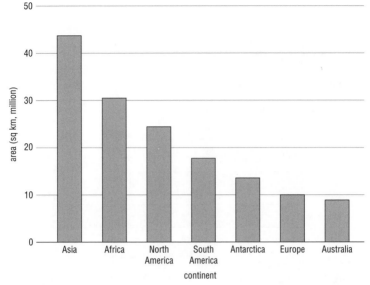

Areas of world continents

Figure 3.2 Re-ordered bar chart
Source: Table 3.2.

The bars of a bar chart usually run vertically in columns, as with Figures 3.1 and 3.2 . In fact, if you are creating these sorts of charts using a spreadsheet or computer graphing package you will probably find that vertical bar charts are more correctly referred to as *column charts*.

In situations where the names are rather wordy, there may not be enough space to write them along the horizontal axis. In this case, it would make sense to draw the bars horizontally rather than vertically, as shown in Figure 3.3. With a spreadsheet or graphing package, such a chart may simply be referred to as a *bar chart*.

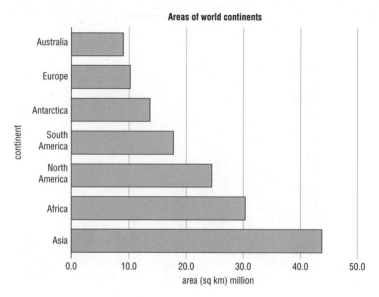

Figure 3.3 Horizontal bar chart

In general, then, horizontal bars are useful when there is quite a large number of categories and some of the category names are long.

Two other forms of bar chart are worth mentioning – the compound and component bar chart.

A *compound bar chart* is a bar chart where the bars are grouped two or three at a time in order to convey information about more than one measure. For example, the compound bar chart in Figure 3.4 shows the average life expectancy in 2014 for a selection of four particular continents. The added dimension here is that the bars are arranged in pairs to show the information for men and women separately. Notice that there is a *key* (sometimes called a *legend*) on the right matching the shading of the bars, which shows which bars refer to men and which to women.

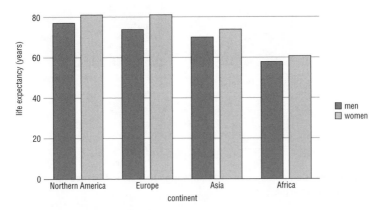

Figure 3.4 Compound bar chart

Source: Statista, The Statistics Portal, www.statista.com/statistics/270861/life-expectancy-by-continent/

The main feature of this bar chart is that it enables the reader to interpret two different sets of information – for example it reveals that women have a greater life expectancy than men but also that the continent of Africa has a lower life expectancy than other continents.

A *component bar chart* can be drawn with each bar split up to reveal its component parts. This is known as a component bar chart but it is also known (for obvious reasons) as a 'stacked' bar chart. Figure 3.5 shows a component bar chart comparing the four most popular media at an annual art exhibition this year and ten years earlier. The data are shown in Table 3.3.

Table 3.3 Artists exhibiting in the four most popular media (%)

Media	This year (%)	Ten years ago (%)
Painting	52	75
Ceramics	26	10
Textiles	13	12
Jewellery	9	3

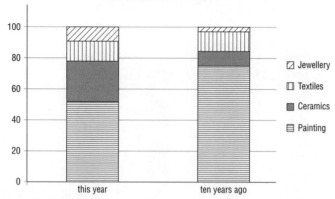

Figure 3.5 Component bar chart showing changes in the media, over time, of an art exhibition

As can be seen from the chart, there has been a change, over time, in the composition of the exhibiting artists. Ten years ago, painting was very much the dominant medium whereas this year the other media are better represented.

In this example, four categories have been used for each bar. As you can imagine, readability would be compromised if too many components were used in the bars. There are no hard and fast rules on the recommended number of components to use, but a maximum of five or six seems sensible.

Exercise 3.1 Revision on bar charts

Here are a few questions to help you clarify for yourself some of the main points about bar charts.

a Why are bar charts drawn so that the bars are all the same width?

b Why are bar charts drawn so that there are gaps between adjacent bars?

c If there is no natural sequence to the categories on the horizontal axis of a bar chart, what would be a sensible basis on which to order the bars?

d Distinguish between a compound bar chart and a component bar chart.

[Comments at the end of chapter]

Pie charts

Figure 3.6 shows the data from the 'This year' column of Table 3.3 represented in a pie chart. As you can see, a pie chart represents the component parts that go to make up the complete 'pie' as a set of differently sized slices.

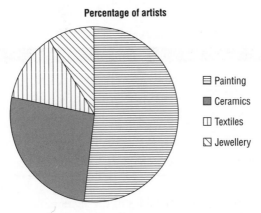

Figure 3.6 Pie chart showing the media displayed at an art exhibition

These days, most people create pie charts on a computer (perhaps on a spreadsheet or a graphing package). Drawing a pie chart by hand requires both care and effort in how the size of each slice is calculated and how it is drawn. The size of each slice as it appears on the pie chart will be determined by the angle it makes at the centre of the pie. You may remember that a full turn is 360° (360 degrees), so the angles at the centre in all the slices must add up to 360°. Taking Figure 3.6 as an example, the Jewellery slice should take up 9% of the pie, so the angle at the centre of the pie made by that particular slice must be 9% of the entire 360°. If you have a calculator to hand calculate:

$$360 \times \frac{9}{100} \text{ to get the answer } 32.4°$$

Exercise 3.2 Checking out the slices

a Based on the data in Table 3.3, and using either a calculator or pencil and paper, calculate the angles at the centre for the other three slices of the pie in Figure 3.6 and record your answers in a table such as the one in Table 3.4.

b Either using a protractor or by eye, check that your answers below correspond to the angles in the slices drawn in Figure 3.6.

Table 3.4 Calculating the angles of the slices

Medium	This year (%)	Angle (°)
Painting	52	
Ceramics	26	
Textiles	13	
Jewellery	9	$360 \times \dfrac{9}{100} = 32.4°$

[Comments at the end of the chapter]

Notice that the numerical values of the slices have not been included in Figure 3.6. However, it is a common practice to do so – sometimes this information will take the form of actual numerical data from the original source and sometimes shown in percentage form.

Pictograms

Pictograms (also known as pictographs or ideographs) are really bar charts consisting of little pictures which indicate what is being represented. A simple example is shown in Figure 3.7.

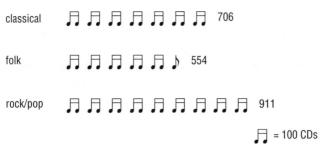

classical   706

folk   554

rock/pop   911

= 100 CDs

Figure 3.7 A pictogram showing the various numbers of CDs in a library, by type
Source: personal survey.

One or two aspects of the pictogram are worth clarifying. First, notice the key at the bottom, which explains that one double-note symbol corresponds to 100 CDs. Also, you may have spotted that the final symbol in the Folk row has been 'cut in half' to indicate that the number of CDs in this category lies between 500 and 600. One final point: particularly where each line of the pictogram consists of several symbols, it is important to ensure that each symbol is the same size. The reason for this is to ensure that the length of each line of pictures is a consistent measure of the number of items it represents and that, visually, you are comparing like with like.

Histograms

The histogram in Figure 3.8 shows the proportions of babies born to mothers of different ages in a typical UK hospital.

A key distinction between the histogram in Figure 3.8 and the bar chart described earlier is that here the adjacent bars are made to touch. The reason for this is to emphasize the fact that the horizontal axis on a histogram represents a continuous number scale. It is important to be aware that the bars on a histogram do not represent separate categories, as they do on a bar chart, but, rather, adjacent intervals on a number line. In other words, although superficially the bar chart and the histogram look similar, they are designed to represent two quite different types of data. The bar chart is useful for depicting separate *categories*, while the histogram describes the 'shape' of data that have been *measured on a continuous number scale*. The distinction between these two types of data is an important one and will be returned to in the next chapter. In the example above, the data have been grouped into intervals of five years on the horizontal axis. This is known as the class interval and in Figure 3.8 the class intervals are all equal.

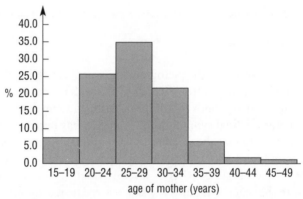

Figure 3.8 Histogram showing the number of live births by age of mother
Source: estimated from national data.

A further subtlety of a histogram which distinguishes it from the bar chart is that the widths of its bars (which correspond to the class intervals) need not all be the same. They may differ if the original data were collected into unequal class intervals in the first place or perhaps if you chose to collapse two or more adjacent intervals together. Suppose, for example, you wish to redraw Figure 3.8, collapsing together all the women aged over 30 years into one class interval. This would produce the frequency table shown in Table 3.5.

Table 3.5 Frequency table showing four adjacent class intervals collapsed together

Age of mother (years)	Frequency (%)
15–19	8.2
20–24	26.9
25–29	35.4
30–49	29.6

Exercise 3.3 Spot the deliberate mistake

Figure 3.9 shows an incorrect histogram drawn from the data in Table 3.5. What do you think is wrong with the way it has been drawn and how would you correct it?

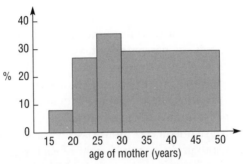

Figure 3.9 Incorrect histogram drawn from Table 3.5

[Comments in the text]

Comparing the overall shape of Figure 3.9 with that of Figure 3.8, it seems that the process of collapsing several class intervals together has the misleading effect of accentuating the importance of the wider interval. Since this interval is four times as wide as the other intervals, it makes sense to adjust for this by reducing the height of this bar to a quarter of the height shown in Figure 3.9, as in Figure 3.10.

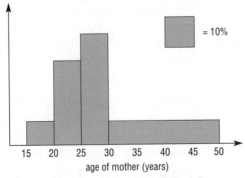

Figure 3.10 Correct histogram drawn from Table 3.5

In effect, the height of the final wide column in Figure 3.10 is an average of the corresponding four separate bars in Figure 3.8 which it has replaced. What this means is that the total area for each of these histograms is the same – an important property of histograms and one which acts as a useful guiding principle if you ever wish to redraw them using different class intervals.

Finally, let's turn our attention to the scale on the vertical axis, which shows the percentage of women contained in each age interval. Where all the class intervals are the same width, as is the case in Figure 3.8, it seems to be reasonably clear what these percentage figures refer to. However, for histograms such as Figure 3.10, where the class intervals are unequal, a number scale on the vertical axis is less meaningful since the height of each bar must also be interpreted in terms of how wide it is. The key property which can be relied on is the *area* of each bar since, as was mentioned earlier, the total area of a histogram must be preserved no matter how the class intervals are altered. Thus, as is shown in Figure 3.10, it is more correct to draw a histogram without a vertical scale but, instead, with its unit area defined alongside the graph.

Stemplots

Stemplots, sometimes called 'stem-and-leaf' diagrams, can often be used as an alternative to histograms for representing numerical data. Table 3.6, for example, gives some numerical data printed in a national newspaper over a period of several consecutive days. They have been taken from a birthdays column and reveal the ages of 40 women deemed to be sufficiently famous to warrant inclusion in the newspaper.

The stemplot started in Figure 3.11 shows how the first four ages are entered. As you can see, the vertical column on the left-hand side, called the 'stem', represents the 'tens' digit of the numbers. The corresponding 'units' digits appear like leaves placed on the horizontal branches coming out from the stem.

For example, the number 32 appears as a 2 (the units digit of the number) placed on level 3 (the tens digit of the number) of the stem, and so on.

Table 3.6 The ages of 40 'famous' women listed in the birthdays column of a national newspaper

77	32	55	55	59	67	55	60	51	82
66	66	100	29	61	47	52	46	53	63
74	47	58	72	55	50	36	52	58	48
80	41	54	53	70	68	42	62	98	45

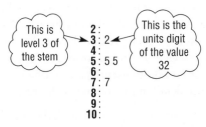

Figure 3.11 Incomplete stemplot representing the first four items from Table 3.6

Exercise 3.4 Completing the stemplot

Draw the stemplot in Figure 3.11 on a piece of paper and then complete it using the data in Table 3.6.

[Comments at the end of the chapter]

The completed stemplot shown in the Comments on exercises section has one important defect in that the leaves have been entered onto each stem in the order in which they appeared in Table 3.6. There is nothing special about this ordering and it makes sense to sort the stemplot so that the 'leaves' on each level are ranked from smallest to largest, reading from left to right. The final version, called a 'sorted' stemplot, is shown in Figure 3.12.

```
 2 : 9
 3 : 26
 4 : 1256778
 5 : 01223345555889
 6 : 01236678
 7 : 0247
 8 : 02
 9 : 8
10 : 0
```

10 : 0 is 100 years

Figure 3.12   'Sorted' stemplot based on the data from Table 3.6

As you can see, the stemplot looks a bit like a histogram on its side.[1] However, it has certain advantages over the histogram, the main one being that the actual values of the raw data from which it has been drawn have been preserved. Not only that but, when the stemplot is drawn in its sorted form, the data are displayed in rank order. These and other useful features of a stemplot will be looked at in more detail in the following chapter.

When two sets of data are to be compared, they can be drawn on two separate stemplots and placed back-to-back. Figure 3.13 shows a back-to-back stemplot to compare the age profiles of successful men and successful women, as represented by their inclusion in the birthdays columns of the same national newspaper.

There are interesting patterns evident in this particular back-to-back stemplot, but we will defer speculating about them until the next chapter.

Before leaving the stemplot, it is worth noting that the example chosen so far to illustrate its use has used data which are two-digit numbers. With values like these, it is fairly obvious that the stem should represent 'tens' and the leaves should correspond to 'units'. However, if the data to be displayed were all decimal numbers less than 1 or very large numbers bigger than, say, 1000, then the stem and units would need to be redefined accordingly. In doing so, you may need to round the data to two significant figures, as the stemplot cannot easily cope with displaying numbers containing three or more digits.

---

[1] In Figure 3.12 the stem has been drawn so that the stem values increase as you read downwards. Some textbooks show them the other way up. The main advantage of drawing them as shown here seems to be that, if rotated through 90°, the stemplot immediately looks like a histogram.

```
      men              women

                2 : 9
          8     3 : 26
      99863     4 : 1256778
8633111110      5 : 01223345555889
  997666442     6 : 01236678
    9764411     7 : 0247
   72211000     8 : 02
                9 : 8
               10 : 0

               10 : 0 is 100 years
```

Figure 3.13 Back-to-back stemplot comparing 'successful' women's and men's ages

Nugget: stemplots

Stemplots are the invention of an American statistician called John Tukey (1915–2000). Around 1972, Tukey found that he was unhappy with the range of charts available to his students for making sense of data quickly and easily. Rather than just moan about it, he invented a variety of new ways of depicting numbers, of which the stemplot is just one. He also invented the boxplot, which is also known as the box-and-whisker plot, and you will learn about this in Chapter 5.

Scattergraphs

Scattergraphs (sometimes known as scatterplots or scatter diagrams) are useful for representing paired data in such a way that you can more easily investigate a possible relationship between the two things being measured. Paired data occur when two distinct characteristics are measured from each item in a sample of items. Table 3.7, for example, contains paired data relating to 12 European countries – the area (in 1000 km^2) of each country and also their estimated populations (in millions) for the year 2100. Figure 3.14 shows the same data plotted as a scattergraph.

Scattergraphs are a particularly important form of representation in statistics as they provide a powerful visual impression of what sort of relationship exists between the two things under consideration. For example, in this instance there seems to be a clear pattern in the distribution of the points, which tends to confirm what you might have already expected from common sense, namely that countries with large areas tend also to have large populations. Understanding what is being revealed by a scattergraph is an important first step in getting to grips with two related statistical ideas which are examined in Chapters 11 and 12: regression and correlation.

Table 3.7 Area and estimated populations of 12 European countries

Country	Area (1000 km²)	Estimated population for 2100 (million)
Belgium	30.5	12.6
Denmark	43.1	6.0
France	544.0	80.3
Germany	357.0	70.4
Greece	132.0	11.1
Ireland	68.9	7.1
Italy	301.3	55.6
Luxembourg	2.6	0.7
Netherlands	41.2	17.4
Portugal	92.1	6.8
Spain	504.8	45.0
United Kingdom	244.1	75.7

Source: The Guardian www.theguardian.com/news/datablog/2011/may/06/world-population-country-un

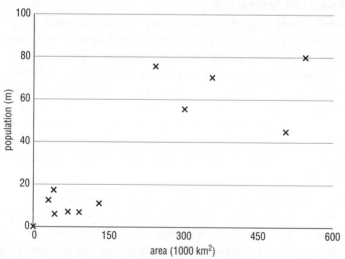

Figure 3.14 Scattergraph showing the relationship between population and area in 12 European countries

Time graphs

Figure 3.15 shows typical time graphs. Time is almost always measured along the horizontal axis and in this case the vertical axis shows the estimated world population at intervals since the 15th century.

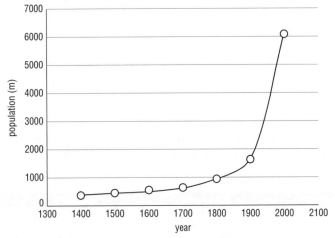

Figure 3.15 Estimate world population, 1400–2000

Source: https://en.wikipedia.org/wiki/World_population_estimates

There is an important difference between time graphs and scattergraphs. With time graphs, each point is recorded in regularly spaced intervals going from left to right along the time axis. Since they therefore form a definite sequence of points, it makes sense to join them together with a line, as has been done here. However, points plotted on a scattergraph are not consecutive and it would be incorrect to attempt to join the points on a scattergraph in this way.

Comments on exercises

▶ **Exercise 3.1**

a Provided that the bars are the same width, the height of each bar gives a fair impression of its value. Where bars are unequal in width, your eye would tend to be drawn to the wider one and, intuitively, you might give it undue weighting.

b The gaps between the bars emphasize that each bar represents a separate category and there is not a continuous number scale on the horizontal axis.

c All things being equal, the most helpful ordering for the bars of a bar chart is usually to start with the tallest bar on the left and place them in descending order from left to right.

d A compound bar chart shows a number of sets of bars grouped together on the same graph – usually in clusters of two or three at a time. With a component bar chart, each bar is drawn with its component parts shown stacked vertically within it.

▶ Exercise 3.2

The table below shows the angles at the centre of each pie (rounded to the nearest whole number) and how they were calculated.

Medium	This year (%)	Angle (°)
Painting	52	$360 \times \dfrac{52}{100} = 187.2°$
Ceramics	26	$360 \times \dfrac{26}{100} = 93.6°$
Textiles	13	$360 \times \dfrac{13}{100} = 46.8°$
Jewellery	9	$360 \times \dfrac{9}{100} = 32.4°$

▶ Exercise 3.4

The complete stemplot looks like this:

```
 2 : 9
 3 : 26
 4 : 7678125
 5 : 55951238502843
 6 : 70661382
 7 : 7420
 8 : 20
 9 : 8
10 : 0
```

Key ideas

This chapter introduced the main features of the following graphs:

▶ bar charts

▶ pie charts

▶ pictograms

▶ histograms

▶ stemplots

▶ scattergraphs

▶ time graphs.

To find out more via a series of videos, please download our free app, Teach Yourself Library, *from the App Store or Google Play.*

4

Choosing a suitable graph

In this chapter you will learn:

▶ *how to distinguish between two important types of data (discrete and continuous)*

▶ *the difference between single and paired data*

▶ *common types of statistical judgement (summarizing, comparing and inter-relating)*

▶ *how to choose the right graph for your needs.*

In the previous chapter, a variety of different types of graph were merely described, without any real explanation of when and how they might be used. We now turn to this question of deciding which graph is most suitable for which situation.

Essentially, two key factors help you to determine which is the best graph to draw. These are the *type of data* being represented and the *type of statistical judgement* which you hope the graph will help you to make. These two aspects are discussed in the first two sections below. In the final section of the chapter, examples will be provided to give you practice at choosing the best graph to represent your data, guided by the principles of the first two sections.

Types of data

Statisticians have come up with a variety of sophisticated and elegant ways of classifying data according to a variety of different attributes. However, since the whole point of this section is to help you decide which graph to choose, we will restrict attention to a simple way of classifying data which informs the choice of graph.

DISCRETE AND CONTINUOUS DATA

The most basic distinction that can be made between types of data is to separate those that are *discrete* from those that are *continuous*. A dictionary definition of the word discrete will read something like 'separate, detached from others, individually distinct, discontinuous'. Table 4.1 shows a few examples of discrete data.

Nugget: discrete, discreet

Don't confuse the words *discrete* and *discreet*; they sound the same but have quite different meanings. Whereas 'discrete' has the meaning explained above, 'discreet' means 'being careful to avoid embarrassing or upsetting others'.

In all of the four examples shown in Table 4.1, each separate item of data collected can be placed into one of a limited number of predefined categories. For example, in the wildflower survey, if the next flower you find is a marigold, then it falls clearly into a 'separate, detached from others, individually distinct' category along with all the other marigolds. Similarly, in the household survey, there is a restricted number of separate household sizes – 1, 2, 3 people, and so on – into which the data can be recorded. Household sizes in between these values, say 2.73 people, or 3.152 people, for example, are simply not possible.

Table 4.1 Examples of discrete data

Investigation	Typical items of data
Types of wildflower	Bluebell, marigold, meadow-sweet, ...
Household size	1 person, 2 people, 3 people, 4 people, ...
Environmental problems	Oil spills, acid rain, dog fouling, ...
Vehicle colour	Red, green, white, ...

As you can see from these examples, discrete data can take the form of either *word categories* (bluebell, acid rain, etc.) or *numbers* (the number of people in a household, for example). However, although word categories are always discrete, this is not always true of data in the form of numbers. Table 4.2 shows examples of numerical data which are not discrete and these form our second fundamental classification – known as *continuous data*.

Table 4.2 Examples of continuous data

Investigation	Typical items of data
Babies' birth weight	3120 g, 3760 g, 2700 g, ...
Temperature survey	18.6°C, 21.4°C, 19.0°C, ...
Survey of commuting times	23 min, 11 min, 70 min, ...

As you can see from these examples, continuous data are all numerical. However, unlike the survey of household size examples earlier, they are not restricted in the number of different values that they can take. Essentially the only restriction here lies in the degree of precision with which the measurements were taken. For example, a survey in which babies' birth weights are measured to the nearest gram will draw on ten times as many possible separate weights as a survey which weighed them to the nearest 10 g. But the point is that, in theory, there is no limit to the number of possible weights at which a baby might be measured and this is a crucial property of continuous data in general.

Two further terms which are often used in this context are *attributes* and *variables*. An attribute is another term for a 'word category' and attributes will generate data that are non-numerical. For example, colour is an attribute and a survey of vehicle colours will produce categorical data – red, green, blue and so on. A variable, on the other hand, is a measure which generates numerical data. Thus, household size and babies' birth weights are both variables, the first producing discrete numerical data (1, 2, 3, etc.) and the second producing continuous numerical data.

Confused? Well, now might be a good time to review what you have read so far.

Exercise 4.1 Consolidation

Have a look now at Table 4.3. Copy and complete the table by ticking the appropriate columns to summarize the distinction between discrete and continuous data, variables and attributes.

Table 4.3 Discrete and continuous, variables and attributes

	Word categories	Discrete numbers	Continuous numbers
Discrete data			
Continuous data			
Variables			
Attributes			

[Comments at the end of the chapter]

It is quite helpful to have, in your mind, an image of continuous data as an infinite number of possible positions on a continuous number scale. Discrete data, on the other hand, whether formed from word categories or discrete numbers, can be thought of as a restricted collection of boxes into which each new item in the sample can be classified.

► **Word categories**

Survey of types of wildflowers

| Bluebells | Marigolds | Meadow-sweets |

► **Discrete numbers**

Survey of household size | 2 people | 3 people | 4 people |

► **Continuous numbers**

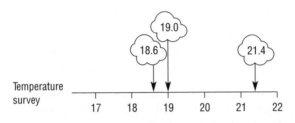

Temperature survey

Later in the chapter a clear link will be made between the type of data you are dealing with (whether they are discrete or continuous) and the most suitable graph to draw, so it is important to be able to distinguish discrete from continuous measures. Exercise 4.2 will allow you to get some practice at this, so do this exercise now.

Although certain variables, such as length, weight, time and so on, can be described as 'continuous', truly continuous measurement is never attainable in practice. The reason for this is that there is no such thing as perfect precision in the real world. For example, if you need to measure a spark plug gap or the thickness of a human hair, even with access to the world's most accurate ruler you are eventually going to have to round your answer to a certain number of decimal places. So, in a sense, all practical measurement is discrete.

Nugget: measuring A4 paper

A rather useful fact that I carry around in my head is that a sheet of A4 paper is 210 mm wide. Do I mean exactly 210 mm? Well, actually, no! I recently viewed a sheet of A4 paper at × 800 magnification using an electron microscope and was amazed to find that, rather than the edges of the paper being perfectly straight, they were very ragged indeed. So, if asked to measure the width of a sheet of A4 paper to, say, the nearest micrometre (1 micrometre is one thousandth of a millimetre or, to put it another way, one millionth of a metre), you just couldn't do it as you wouldn't know precisely where the paper width started and ended!

Exercise 4.2 Discrete or continuous?

In the extract below, a number of items of data have been written in bold type. Try to sort out which are discrete and which are continuous and record your answers in a table such as Table 4.4.

The dwarfing apple rootstock 'M27' was raised in **1929** from a cross between 'M9' and 'M13'. As a dwarf bush, it makes a tree **1.2–1.5 m** in height and spread. A well-grown tree should yield on average **4.5–6.8 kg** of fruit each year. At planting, side-shoots are cut back to **three buds** and the leader pruned by about **one quarter**, cutting to an upward facing bud.

Adapted from Harry Baker, *Stepover Apples*, The Garden, Volume 116

Table 4.4 Classifying discrete and continuous data

Data item	Type of data (D or C)
M27	
1929	
1.2–1.5 m	
4.5–6.8 kg	
Three buds	
One quarter	

[Comments at the end of the chapter]

SINGLE AND PAIRED DATA

As was indicated in the section dealing with scattergraphs in the previous chapter, data can be collected either *singly* or *in pairs*. The meaning of these two terms is quite straightforward. If only one measure is taken from each item in the sample, the data have been collected singly. If two different characteristics are measured from each data item, then the data are said to be 'paired'. Here are a few examples of each.

▶ **Single data measurement**

(a) Weights of 30 eggs (nearest gram): 66, 73, 54, 59, 61, ...

(b) Temperature over 30 days (°C): 19, 17, 15, 18, 21, 14, ...

(c) Type of wildflowers collected: bluebell, marigold, meadow-sweet...

▶ **Paired data measurement**

Weights and sizes of 30 eggs (nearest gram)

Egg number	1	2	3	4	5	...
Weight (g)	66	73	54	59	61	...
Size	2	1	5	4	3	...

Average temperature (°C) and rainfall (cm) over 12 months

Month	Jan	Feb	Mar	Apr	May	...
Temperature (°C)	7	9	11	12	16	...
Rainfall (mm)	170	140	110	90	80	...

Paired data are described more fully in Chapter 11.

Types of statistical judgement

Data are normally collected and graphed for a particular purpose: to help us make a statistical judgement. In Chapter 1, three of the most common types of statistical judgement that are made were described, and these are still fundamental to helping you decide what sort of graph you should draw. First, graphs are often useful simply to *summarize* the data. Graphs are also useful for *comparing* and *inter-relating* and we will look at each of these three types of judgement in turn.

SUMMARIZING

The most basic purpose of a graph is to summarize the essential characteristics of the data in question. You may be wondering, however, what exactly are the key characteristics that are worth examining, so let us start by having a look at the data in Figure 4.1, which have been taken from the previous chapter.

```
 2 : 9
 3 : 26
 4 : 1256778
 5 : 01223345555889
 6 : 01236678
 7 : 0247
 8 : 02
 9 : 8
10 : 0
```

10 : 0 is 100 years

Figure 4.1 Stemplot showing the ages of 40 'famous' women
Source: Figure 3.12.

Exercise 4.3 The bare essentials

Jot down two or three of the essential features of these data that are revealed by the way they have been presented in this stemplot.

[Comments below]

Two key features are worth stressing here, both of which are ways of summarizing the data. The first is that a typical or average or 'central' value is around 55 years. This is a description of the *central tendency* of the data – a rather pompous statistician's way of describing where the data are centred. The second useful summary is to describe *how widely spread* the data are. There are several ways of doing this, the easiest of which is just to state the lowest and the highest values. So, in this example, the range is 71 years, running from a low of 29 to a high of 100 years.

To recap, then, the two most common ways of summarizing data are, first, using a measure of central tendency (perhaps an average) and, second, by measuring the spread. Formal methods for performing these calculations are described in Chapter 5, but for the purposes of this chapter it is enough to be able to have an intuitive sense of them simply by looking at a graph.

COMPARING

A fundamental question in statistics is whether one set of measurements is longer, heavier, warmer, longer-lasting, healthier, better value for money, etc. than another. There is a variety of statistical tests designed to measure just how significant such differences are but these are not covered here as this aspect of statistics is beyond the scope of this book. However, Chapter 14, 'Deciding on differences', will give you a common sense, intuitive understanding of this important topic. Whether or not you decide to study statistical tests of significance in more detail, it would be a good idea to develop an intuition about the idea of making such comparisons by purely graphical methods. Figure 4.2 shows a back-to-back stemplot comparing 'successful' women's and men's ages. (This was given in the previous chapter as Figure 3.13.)

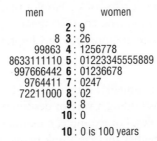

```
        men                women
                      2 : 9
                  8   3 : 26
              99863   4 : 1256778
       8633111110   5 : 01223345555889
       997666442   6 : 01236678
         9764411   7 : 0247
        72211000   8 : 02
                      9 : 8
                     10 : 0

                     10 : 0 is 100 years
```

Figure 4.2 Back-to-back stemplot comparing 'successful' women's and men's ages
Source: Figure 3.13.

Exercise 4.4 Interpreting a back-to-back stemplot

a How does the back-to-back stemplot in Figure 4.2 help you to make a comparison between the two batches of data?

b What additional calculations might help you to make this comparison more fully?

[Comments below]

a It is clearly evident from Figure 4.2 that the women's ages are more widely spread than those of the men (a range of 29–100 years, as opposed to 38–87 years for the men). Also, the women were, overall, slightly younger than the men. (15 men were in the '70 and over' category, as opposed to only eight women.)

b While this back-to-back stemplot may give you a reasonably good intuitive feeling for the differences between the two batches of data, it would also be helpful in this case to calculate the averages of the women's and men's ages. The most common form of average is the mean – found by adding all the values in the batch together and dividing by the number of values in the batch. The mean age of the men in Figure 4.2 turns out to be 64.2 years, while the mean age of the women is only 58.5 years, confirming the impression conveyed by the back-to-back stemplot that the men are rather older than the women. (The mean and other types of average are explained more fully in Chapter 5.)

INTER-RELATING

The third of our fundamental statistical judgements is inter-relating. This means finding a way of describing the relationship between two variables. For example, a manufacturer might be interested to see whether there is a link between sales and the amount of money spent on advertising. A scientist may wish to explore the connection between the temperature of a metal bar and how long it is. An important relationship in medicine is the connection between the drug dosage and its effects on patients. Environmentalists may want to investigate the link between the burning of fossil fuels and resulting changes in the weather. In each of these cases, there are not one but two separate variables being considered and it is their inter-relationship which is of particular interest. The most likely approach will be to collect and analyse paired data taken from each of the variables together. For example, Table 4.5 provides paired data from which you might be able to investigate a possible inter-relationship between someone's height at age 2 years and their eventual adult height.

Table 4.5 Paired data showing height at 2 years and eventual adult height

Name	Height at age 2 years (m)	Adult height (m)
Alan	0.84	1.71
Hilary	0.80	1.56
Luke	0.89	1.80
Ruth	0.72	1.50
Ian	0.86	1.76
Pauline	0.83	1.65
Bal	0.88	1.72
Marti	0.79	1.60
Francis	0.85	1.73
Iris	0.77	1.55

Source: personal data.

The most helpful type of graph for revealing inter-relating patterns in paired data is the scattergraph.

Getting the right graph

So far in this chapter you have been introduced to two key elements which inform your choice of graph. These are the *type of data* being represented and the *type of statistical judgement* which you hope the graph will help you to make. You may already have seen that these two elements are not entirely separate from each other since, not surprisingly, the sort of data that you decide to collect is bound to be related to the sort of decision you intend to make about them. A clear example of this is with paired data, which you would almost certainly have collected in order to explore an inter-relationship. In this final section some examples are provided which will tie these ideas together with what you have learned about the key features of the common graphs from Chapter 3. You will be asked to look at some of the graphs in Chapter 3 and make an assessment of them in terms of the two key elements described above. In order to simplify the task, this box provides you with a handy summary of the key ideas introduced so far.

Type of data:	discrete	continuous	
Type of judgement:	summarizing	comparing	inter-relating

BAR CHARTS AND PIE CHARTS

Let us start with bar charts and pie charts.

The areas of world continents listed in Table 3.2 and the average life expectancy contained in Figure 3.4 refer to discrete data in both cases. However, the type of statistical judgements which they inform are rather different. Whereas the simple bar chart in Figure 3.1 merely *describes* the relative areas of the world's continents, the compound bar chart in Figure 3.4 is helpful for *comparing* two different batches of data, one for men and the other for women. Now you may have felt that the simple bar chart did enable you to make comparisons – comparing between its various bars. However, this is not the sense in which the term comparing was originally used earlier in the chapter. Comparing, as the term is used here, is concerned with how one set of measurements differs from another.

Exercise 4.5 Analysing bar charts

Have a look at Figures 3.1 and 3.4 and for each of these bar charts use the boxed summary above to identify:

a the type of data being represented

b the type of statistical judgement which they might enable you to make.

[Comments below]

In general, a bar chart should be used only when the quantity on its horizontal axis is *discrete*. This may be a set of categories drawn from an attribute, as in the case of Figure 3.1. Alternatively, the data may be discrete numerical, such as those in the example in Figure 4.3. The types of statistical judgement are, respectively, *describing* and *comparing*.

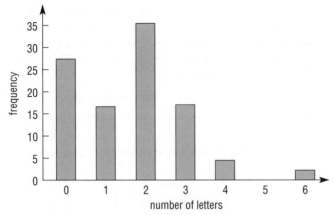

Figure 4.3 A bar chart used to represent discrete numerical data – the number of letters delivered to the same address
Source: personal survey.

If you wish to compare two batches of discrete data, then use two separate bar charts, a compound bar chart or a component bar chart.

We now turn to pie charts.

Exercise 4.6 Analysing pie charts

Have a look at the pie chart in Figure 3.6 and try to identify:

a the type of data being represented

b the type of statistical judgement which it might enable you to make.

[Comments below]

Your responses should have been the same as those for the bar chart in Figure 3.1, namely that the data being represented are *discrete* and the judgement is one of *describing*. As before, a case could be made for claiming that the pie chart allows you to make comparisons between the slices of the pie. However, this is not the sense in which the term 'comparing' is being used here.

In general, like the bar chart, a pie chart should be used only to depict discrete data. These may be a set of categories drawn from an attribute, as in the case of Figure 3.1. However, unlike for the bar chart, it does not make sense to draw a pie chart for data which are discrete numerical.

Exercise 4.7 A rather half-baked pie chart?

Have a look at the example in Figure 4.4, and try to decide why this pie chart may not be the ideal representation for these data.

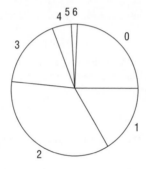

Figure 4.4 Inappropriate use of a pie chart, showing the number of letters in 100 deliveries

[Comments below]

Although the pie chart in Figure 4.4 isn't entirely wrong, it has two rather silly features which you may have spotted. First, it doesn't make much sense to place a number scale in a circular arrangement, as has been done here. The numbers 0 to 6 fall in a natural sequence and it would be more sensible to find a representation which preserves and even emphasizes this feature – a straight line would be best. Second, you may remember from the bar chart in Figure 4.3 that the bar corresponding to the category '5 letters' had a frequency of zero. Unfortunately, in a pie chart, any category with a frequency of zero simply disappears. The bar chart has neither of these drawbacks.

However, pie charts do have a special quality which simple bar charts do not.[1] This is that they give a good impression of how

---

[1] Component bar charts do, however, meet this condition.

large each 'slice' is in relation to the complete pie. In fact, a pie chart should not be used unless the complete pie meaningfully represents the sum of the separate slices. In the case of the artists data (Figure 3.6) the complete pie corresponds to all the different artistic mediums taken together, so it does meet this condition. However, Figure 4.5 shows a pie chart where the condition is not met, and as a result the complete pie chart is meaningless. It has been drawn from the data in Table 4.6.

Table 4.6 Equatorial diameter of the four largest planets

	Diameter (1000 km)	
Jupiter	143	
Saturn	121	
Uranus	51	
Neptune	50	

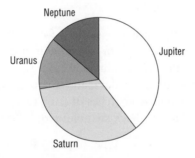

Figure 4.5 A meaningless pie chart drawn from the data in Table 4.6.

STEMPLOTS AND HISTOGRAMS

We now turn our attention to stemplots and histograms.

Exercise 4.8 Analysing stemplots

Have a look at Figures 3.10 and 3.13 and for each of these representations try to identify:

a the type of data being represented

b the type of statistical judgement which they might enable you to make.

[Comments below]

a The data represented in Figure 3.10 were taken from Table 3.5. Table 3.5 is itself a summary of the original raw data, so we have to use our imaginations a little to decide what the original data might have looked like. It is likely that they were a batch of ages, probably measured in whole years. Since age is a continuous variable, it would be reasonable to describe these data as continuous.

However, as was indicated earlier, the data were probably collected in discrete numbers of years, so there is room for ambiguity in this answer. The stemplot data in Figure 3.13 are also taken from the same continuous variable (age), but again it is clear that they have been stated in discrete numbers of years.

b Whereas the histogram simply describes the data, the back-to-back stemplot in Figure 3.13 allows you to compare two batches of data side by side.

In general, histograms are only suitable for continuous data, while stemplots can be used for any type of numerical data – discrete or continuous. Individual histograms and stemplots are useful for describing the data, and if you need to make comparisons between two batches of data, then a pair of histograms or a back-to-back stemplot would be needed.

We now turn to the final two types of graph shown in Figures 3.14 and 3.15. These are the scattergraph and the line graph. If you look at these two graphs now, you can see that, in both cases, they have been drawn with two separate variables marked on the axes. It has already been mentioned that these types of graph are useful for representing paired data and allow us to explore inter-relationships.

This completes the description of the common graphs in terms of the sorts of data which they are used to represent, and the sorts of statistical judgements which they enable you to make. The following final exercise in the chapter will allow you to review and summarize these ideas for yourself.

Exercise 4.9 Summary time

Copy and complete Table 4.7. This involves selecting the most appropriate graphs, from the list below, for analysing each type of data. Bear in mind the type of judgements for which each graph is usually used.

List of graphs: bar chart, pair of bar charts, compound bar chart, component bar chart, pie chart, pair of pie charts, pictogram, histogram, pair of histograms, stemplot, back-to-back stemplot, scattergraph, time graph.

Table 4.7 Type of judgement and type of data together

Type of data	Describing	Comparing	Inter-relating
Discrete single data			
Continuous single data			
Paired data			

[Comments at the end of the chapter]

Comments on exercises

▶ Exercise 4.1

	Word categories	Discrete numbers	Continuous numbers
Discrete data	✓	✓	
Continuous data			✓
Variables		✓	✓
Attributes	✓		

▶ **Exercise 4.2**

Data item	Type of data (D or C)
M27	D (word category)
1929*	C
1.2–1.5 m	C
4.5–6.8 kg	C
Three buds	D (discrete number)
One quarter	C

* The year, 1929, is the only item of data in this table which is difficult to classify unambiguously. On the one hand what we have here is a measure of time, in years, and time is a classic textbook example of a continuous measure. However, dates are normally measured in whole numbers of years (we would never talk of the year 1929½, for example) so there would also be a case for calling this a discrete measure.

▶ **Exercise 4.9**

| Type of data | Type of judgement | | |
	Describing	Comparing	Inter-relating
Discrete single data	Bar chart, pie chart, pictogram, stemplot	Pair of bar charts, pair of pie charts, compound bar chart, component bar chart, back-to-back stemplot	
Continuous single data	Histogram, stemplot	Pair of histograms, back-to-back stemplot	
Paired data			Scattergraph, time graph

Note that the stemplot has been placed in both the discrete and the continuous categories in the table above. However, it would only make sense to use a stemplot with a continuous measure if the numbers involved were suitably rounded.

Key ideas

This chapter has built on the technical description of how to draw the various common graphs given in the previous chapter, and has turned to look at when they might usefully be drawn. As described here, the choice of graph depends on two crucial factors:

1 the nature of the data being represented (whether they are discrete, continuous, or paired)

2 the sort of statistical judgement that you wish to make about the data (whether you are describing, comparing or inter-relating).

The final table, which you completed in Exercise 4.9 (and which is completed for you in the Comments on exercises), should provide you with a useful summary of all these ideas in a single diagram.

To find out more via a series of videos, please download our free app, Teach Yourself Library, *from the App Store or Google Play.*

Summarizing data

In this chapter you will learn:

▶ *how to calculate three types of average (mean, mode and median)*

▶ *how to recognize which average is appropriate for which situation*

▶ *measures of spread and how to represent them on a boxplot.*

We live in a world of rapidly growing collections of data. Information is being amassed and communicated on an ever-wider range of human activities and interests. Even if supported by a computer database, a calculator or a statistical package, it is hard for the 'average person' to gain a clear sense of what this information might be telling them. A crucial human skill is to be selective about the data that we choose to analyse and, where possible, to summarize the information as briefly and usefully as possible. In practice, the two most useful questions to help you to summarize a mass of figures are:

What is a typical, or average, value?

and

How widely spread are the figures?

This chapter looks at a few of the most useful summary measures under these two headings: average and spread.

Nugget: only average?

Despite the fact that an 'average' is taken to refer to a typical or representative value, many people use the word average in a pejorative sense. For example, no parent wants their child's school performance to be described as 'just average' and most drivers believe that their driving skills are 'better than average'. If you think about it, it is impossible for all children or all drivers to be better than average. It reflects the fact that parents tend to think that 'all their geese are swans' and that drivers tend to have an over-inflated sense of their own abilities!

Introducing averages – finding a middle value

The next section looks in detail at three particular averages and how they are calculated. First, however, it is worth focusing on what an average is and why we might bother to calculate it. As was suggested earlier, it is often difficult when looking at a mass of figures to 'see the wood for the trees'. A sensible strategy is to find some typical or middle value which may be taken as being representative of the rest. However, there are several possible ways of finding a representative value, as the following simple exercise illustrates. Try Exercise 5.1 now.

Exercise 5.1 Typical of kids

Table 5.1 gives the names of nine female friends and the number of children they have.

Table 5.1 Numbers of children of nine women

Person	Children
Roberta	0
Alice	3
Rajvinder	0
Jenny	1
Fiona	2
Sumita	2
Hilary	3
Marcie	0
Sue	4

How would you summarize this information with a single representative number? See if you can think of several plausible ways of doing this.

[Comments in the text]

Here are three possible ways of summarizing the numbers.

▶ One possible approach that you may have chosen is to ask what is the most common number of children. This can be quickly checked in your head, and your thinking might have involved reorganizing the data as follows.

Number of children	0	1	2	3	4
Frequency	3	1	2	2	1

▷ This reveals that the most frequently occurring number of children is zero (corresponding to a frequency of three), so one possible single summary is:

Average = 0.

▷ This sort of average, the most frequently occurring number, is called the *mode*.

▶ A difficulty with using the mode in this example is that it has come up with a value which could not be described as 'typical' in the sense of lying somewhere in the middle of the range of possible values from 0 to 4. One way of achieving a middle value is by adding the nine numbers together and dividing by nine, thus:

$$\frac{0+3+0+1+2+2+3+0+4}{9} = \frac{15}{9} = 1.667 \text{ children}$$

▶ This type of average is often called the 'average', but clearly this is not a sufficiently precise term since there are several types of average around. The 'correct' statistical term for this type of average is the *mean*, also known as the *arithmetic mean* (pronounced arith-*met*-tic).

▶ The mean has at least given us a representative value which lies within the range of possible values from 0 to 4. However, since children are usually measured in whole units, it has produced a number which can hardly be described as typical! This example seems to demand that we try to find a representative value which both lies within the range 0 to 4 *and* also gives a whole number. Fortunately there is a measure that does just that. What we need to do is to sort the nine numbers into order from smallest to largest and then simply select the middle number in the sequence. For nine numbers, the middle one will be the fifth (giving four numbers on either side). Thus:

Numbers in order of size   0 0 0 1 2 2 3 3 4

This is the middle one

By this method, we get a representative value of two children. This type of average is called the *median*. When there is an odd number of items, as there is here, it is clear that there will be a single median value. However, suppose that there had been an even number of values. For example, let us exclude Sue's four children from the group and try to find the median number of children from these eight numbers.

Numbers in order of size   0 0 0 1 2 2 3 3

These are the middle two

There are now two middle values – the fourth and fifth numbers. In order to find the median, we must now find the mean of these two middle values.

Thus, in this case, median = $\dfrac{1+2}{2}$ or 1.5

In this particular example, the median would seem to be the most suitable sort of average. However, it has to be admitted that this has been a fairly artificial example. Since the choice of average really depends on context and on the purpose for calculating it in the first place, this wouldn't be a good illustration of how you might go about selecting a suitable average for a given problem. Some specific suggestions are given in the next section.

Nugget: weather averages

I have just been listening to Radio 4 where the subject under discussion was that the summer weather was not as the Met Office had predicted. The presenter commented that, in June, July and August of this year, there was a 50% chance of the weather being better than average. It made me wonder what the percentage is during the rest of the year...

Calculating and choosing averages

We now look in a bit more detail at how these three averages – the mode, the mean and the median – are calculated and also at how we can decide which one to choose under which circumstances.

MODE

The mode of a batch of data is usually defined as the most frequently occurring item. The procedure for finding the mode is as follows:

  i identify all the distinct values or categories in the batch of data;

 iii make a tally of the number of occurrences of each value or category; and

 iii the mode is the value or category with the greatest frequency.

Do Exercise 5.2 now, which contains three simple examples to give you practice at calculating the mode for different types of data.

Exercise 5.2 Finding the mode

Three additional sets of data are supplied in Table 5.2, all associated with the nine friends listed in Table 5.1. Respectively, these data sets can be described as

(a) category data (eye colour), (b) discrete numerical data (lucky number) and (c) continuous numerical data (height).

Calculate the mode for each of these data sets.

Table 5.2 Three different data types

Name	(a) Eye colour	(b) Lucky number	(c) Height (m)
Roberta	blue	3	1.56
Alice	green	7	1.62
Rajvinder	brown	7	1.71
Jenny	brown	9	1.58
Fiona	blue	3	1.67
Sumita	brown	5	1.66
Hilary	green	7	1.60
Marcie	brown	7	1.59
Sue	blue	7	1.61

These are category data

These are discrete numerical data

These are continuous numerical data

[Comments at the end of the chapter]

There is a problem when calculating the mode from continuous numerical data, of the type in column (c) of Table 5.2. The difficulty, and one that is in evidence here, is that each number is different. As a result, they each have a tally of one, which makes the mode – the most frequently occurring value – rather meaningless. The problem can be solved by grouping the nine values into intervals of 5 cm, as shown in Table 5.3.

When the data have been grouped, it is possible to point to the interval with the greatest frequency and this is called the 'modal interval' – in this case the modal interval is $1.55 < H \leq 1.60$, with a frequency of four. By the way, the symbol < means 'less than', while the symbol $\leq$ means 'less than or equal to'.

Table 5.3 Tally of heights in 5-cm intervals

Height (H) m	Tally
$1.55 < H \leq 1.60$	IIII
$1.60 < H \leq 1.65$	III
$1.65 < H \leq 1.70$	II
$1.70 < H \leq 1.75$	I

This is the modal interval ...

... because this is the largest frequency

Overall, the mode is particularly useful for categorical data or discrete numerical data but can only be used meaningfully with continuous data if they are grouped into intervals, as has been done here.

A SIMPLE MEAN ($\overline{X}$)

The mean, or the arithmetic mean, as it is sometimes called, is the best known average and can be defined as 'the sum of the values divided by the number of values'. The mathematical symbols used to describe the mean were explained in Chapter 2. Here they are again.

The symbol for the mean, pronounced 'X bar'

$$\overline{X} = \frac{\sum X}{n}$$

The Greek letter capital sigma, meaning 'the sum of'

Exercise 5.3 Finding the mean

Try to find the mean of the three sets of data in Table 5.2.

[Comments below]

▶ If you didn't fail to find the mean of data set (a), you should have! It is simply nonsensical to find the mean of categorical data.

▶ It is possible to calculate the mean of the nine numbers in data set (b). This gives:

$$\overline{X} = \frac{3+7+7+9+3+5+7+7+7}{9} = \frac{55}{9} = 6.111\ldots$$

Wow – some lucky number! Clearly, although a mean can be found here, the number you get, 6.11…, is not remotely useful or interesting, so I'm sorry to have wasted your time! Perhaps the lesson to learn from this exercise is that the mean is a somewhat overworked average, and you need to question whether it meets the main criterion of any well-chosen average, namely does it provide a useful and representative summary of the data set? If it doesn't, then don't use it.

▶ Continuous data of the type given in data set (c) are usually ideal for calculation of the mean. The calculation in this case is:

$$\overline{X} = \frac{1.56+1.62+1.71+1.58+1.67+1.66+1.60+1.59+1.61}{9}$$

$$= \frac{14.60}{9} = 1.622 \text{ m}$$

Note that it is usual to give the answer to the calculation of a mean to one place of decimals more than the original data, as has been done here (three decimal places compared with two in the original heights).

A WEIGHTED MEAN ($\overline{X}$)

We shall now look at what is involved when calculating means of data sets with much larger sample sizes and where the data are organized either in frequency tables or percentage tables, depending on which is the more appropriate. For example,

Table 5.4 shows the distribution of households by size for 1961 and 'today' for a large city in the UK. It simply wouldn't make sense to write each number out separately for every person surveyed. Instead, as shown in Table 5.4, the data have been grouped and, for convenience, are stated in percentages. The figures for 1961 and 'today' are given side by side.

Table 5.4 Households by size (percentage) in a large city in the UK

Household size	1961	'Today'
1	12	25
2	30	34
3	23	17
4	19	16
5	9	6
6+	7	2

Source: adapted from national data.

Seeing the 1961 and 'today' distributions of household sizes side by side makes for an interesting comparison. Two significant differences between these two distributions occur at the extremes, i.e. at household size 1 and 6+. The proportion of single-person households 'today' has roughly doubled compared with 1961 (25% compared with 12%), while the proportion of large households containing 6+ people dropped from 7% to 2%. Overall, then, there has been a significant shift to smaller household sizes and this should be reflected in a reduction in the average household size, as the calculation of the mean below will show.

With this example, the basic definition of the mean as 'the sum of the values divided by the number of values' will not give the correct answer. The particular complication here is that the different household sizes, 1, 2, 3, etc., do not occur with the same relative frequency. Taking the 1961 figures, for example, 12% of the households were of size 1, whereas 30% were of size 2. So clearly these different-sized households are not equally weighted and any averaging procedure would need to take account of the unequal weights. A second complication occurs with the final household size, 6+. This could include households with 7, 8, 9, etc. people, right up to 20 or 50 or maybe even 100! Of course, there will be many more households of size 6 or 7 than there will be, say, of size 20. What we have here, then, is an interval where the upper limit has not been specified. What we need to do is take a sensible number of people, say 9, which would be representative of this interval. So, how do we come to the value 9? Well, without further information, the precise value chosen is just a matter of common sense and intelligent guesswork.

We shall now calculate the mean household size for 1961. This time we shall use the procedure for calculating a 'weighted mean', which involves the following stages:

i Multiply each household size (X) by its corresponding 'weight' or frequency (f). In this example, the f refers to the percentage of households of household size 1, 2, 3, etc.

ii Add these products together and divide by the sum of the weights.

Table 5.5 Calculating a weighted mean for the 1961 data on household size

Household size (X)	Weight (f)	Product (fX)
1	12	$12 \times 1 = 12$
2	30	$30 \times 2 = 60$
3	23	$23 \times 3 = 69$
4	19	$19 \times 4 = 76$
5	9	$9 \times 5 = 45$
6+ (assume = 9)	7	$7 \times 9 = 63$
TOTAL	100	325

This is the sum of the weights

This is the sum of the fX products

Referring to Table 5.5, weighted mean = $\dfrac{325}{100}$ = 3.25 persons per household.

Exercise 5.4 Calculate a weighted mean yourself

Now you have seen the calculation of the weighted mean for the 1961 figures, use the same approach to calculate the weighted mean for the 'today' figures.

Comparing the two results, what do they suggest about how average household size has altered over the period in question?

[Comments below]

The weighted mean for 'today' is 2.56 people per household, which, as we predicted, is substantially less than the 1961 figure of 3.25.

Finally, here is a formal mathematical description of how a weighted mean is calculated.

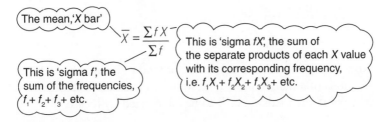

The mean, 'X bar'

$$\bar{X} = \frac{\sum f X}{\sum f}$$

This is 'sigma fX', the sum of the separate products of each X value with its corresponding frequency, i.e. $f_1X_1 + f_2X_2 + f_3X_3 +$ etc.

This is 'sigma f', the sum of the frequencies, $f_1 + f_2 + f_3 +$ etc.

MEDIAN (Md)

As was shown earlier, the median of a set of, say, nine values is found by ranking the values in order of size and choosing the middle one – in the case of nine values it is the value of the fifth one. Note that a common mistake is to say that the value of the median of nine items is five. It needs to be stressed that the median is *the value* of item number five when all nine items are ranked in order and is not its rank number. With an odd number of items, such as 11 or 37 or 135, there will always be a unique item in the middle (respectively the 6th, 19th and 68th). A simple way of working this out is to add one to the batch size and divide by two. For example, for a batch of 37 items, $(37 + 1)/2 = 19$, so the median is the value of the 19th item. However, where the batch size of the data is even, there is not a unique item in the middle. In such circumstances, the usual approach is to choose the two middle values and find their mean.

Example

A count of the contents of eight boxes of matches produced the following results:

Number of matches 52 49 47 55 54 51 50 50

Find the median number of matches in a box.

Sorting the numbers in order of size produces the following:

47 49 50 50 51 52 54 55

(The two middle values)

Since there is an even number of values, the median is the mean of the two middle values, i.e.

$$\text{median} = \frac{50+51}{2} = 50.5$$

Now clearly boxes of matches do not tend to be bought with fractions of a match in the box, so in this example, and indeed in general, the median needs to be interpreted with care.

Let us now look at how we calculate the median from a frequency or percentage table. The stages are as follows:

i Redraw the table as a cumulative percentage table.

ii Identify the value corresponding to the 50% position (remember that the median can be found 50% of the way along the values when they have been ranked in order of size).

Example Find the median household size for the 1961 data in Table 5.4.

Solution Table 5.6 shows the data extended to include a cumulative percentage table.

Table 5.6 Cumulative percentage table showing the 1961 data on household size

Household size	Percentage	Household size	Cumulative percentage
1	12	1	12
2	30	2 or less	42
3	23	**3 or less**	65
4	19	4 or less	84
5	9	5 or less	93
6+	7	all	100
TOTAL	100		

65 = 12 + 30 + 23

The bold line in the cumulative percentage part of Table 5.6, corresponding to a household size of '3 or less', covers the 23 percentage points ranging from 43% to 65% inclusive. Clearly the 50% value lies in this range, so '3' must be the household size that contains the median.

Exercise 5.5 Calculating a median yourself

Now that you have seen the calculation of the median for the 1961 figures, use the same approach to calculate the median for the 'today' figures (also given in Table 5.4).

Comparing the two results, what do they suggest about how average household size has altered over the period in question?

[Comments at the end of the chapter]

Finally, to end this section on averages, Table 5.7 provides a summary of the main characteristics of the three averages and the type of data they can usefully summarize.

Table 5.7 An average for all data types

Type of average	Type of data		
	Category	Discrete numerical	Continuous numerical
Mode	Ideal – only the mode will do for most category data	Fine, provided some of the values occur more than once	Only if the data are grouped into intervals
Mean	No good	Fine, but be careful to check that the answer is sensible	Good
Median	Only if the data have a natural order	Fine, but be careful to check that the answer is sensible	Good

Nugget: mode, mean and median can be muddling

Students often mix up these three averages: mode, mean and median. Here are some tips for remembering which is which.

▶ Mode: the **MO**de is the **MO**st – i.e. the number that shows up the most.

▶ Mean: the mean usually involves a hard calculation and it is **mean** of the teacher to ask this of students!

▶ Median: if you list all three words in alphabetical order (mean, median, mode), the median is the middle term (and the definition of median is that it is the middle value when the numbers are placed in order of size).

Spread

As you have seen in the previous section, knowing the approximate centre of a batch of data is a useful and obvious summary of the entire batch. The second key question which we posed at the beginning of the chapter was, 'How widely spread are the values?' This section deals with the following five measures of spread:

▶ range

▶ inter-quartile range

▶ mean deviation

- variance
- standard deviation.

Since people are increasingly performing statistical calculations like these with the aid of a calculator or suitable computer package (perhaps a spreadsheet or statistical package), treatment of these techniques here will focus less on the mind-numbing arithmetic that can be involved and more on the principles on which they are based. To this end, a very elementary data set has been used throughout, simply to lay out the bare bones of each calculation. The data are the heights of the nine women first given in Table 5.2 and they are repeated in Table 5.8.

Table 5.8 The heights of nine women

Name	Height (m)
Roberta	1.56
Alice	1.62
Rajvinder	1.71
Jenny	1.58
Fiona	1.67
Sumita	1.66
Hilary	1.60
Marcie	1.59
Sue	1.61

FIVE-FIGURE SUMMARY

Before looking at the range and inter-quartile range, it is necessary to scan these nine numbers in order to pick out five of the key figures. This is easiest to do when the numbers are sorted in order of size, as shown in Table 5.9.

- The lower extreme value (E_L) is the height of the shortest person, Roberta, with a height of 1.56 m.

- The upper extreme value (E_U) is the height of the tallest person, Rajvinder, with a height of 1.71 m.

- The median value (Md) lies halfway through the values in Table 5.9. This corresponds to Sue's height, so $Md = 1.61$ m.

- The lower quartile (Q_L) lies a quarter of the way through the values in Table 5.9. To put this another way, the lower quartile lies halfway through the lower half (i.e. covering the five shortest people) of the batch of data. Marcie is third of these five women, so $Q_L = 1.59$ m.

▶ The upper quartile (Q_U) lies three-quarters of the way through the values in Table 5.9. By the same reasoning as before, this will correspond to the median height of the five tallest women. This will be Sumita's height, so $Q_U = 1.66$ m.

Table 5.9 The heights of nine women, sorted in order of size

Name	Height (m)	Five key figures
Roberta	1.56	→ The lower extreme value (E_L)
Jenny	1.58	
Marcie	1.59	→ The lower quartile (Q_L)
Hilary	1.60	
Sue	1.61	→ The median height (Md)
Alice	1.62	
Sumita	1.66	→ The upper quartile (Q_U)
Fiona	1.67	
Rajvinder	1.71	→ The upper extreme value (E_U)

These five numbers provide a useful summary of a batch of data and are often written in the following arrangement, known as a 'five-figure summary'.

$$
n \left|
\begin{array}{cc}
 & Md & \\
Q_L & & Q_U \\
E_L & & E_U
\end{array}
\right.
$$

Included to the left of the five-figure summary is the sample size, 'n', also known as the batch size or the 'count'. The five-figure summary for these data is given in Figure 5.1.

$$
9 \left|
\begin{array}{cc}
 & 1.61 & \\
1.59 & & 1.66 \\
1.56 & & 1.71
\end{array}
\right.
$$

Figure 5.1 Five-figure summary for the heights (m) of nine women

A very effective way of picturing the information in a five-figure summary is to use a boxplot (sometimes called a 'box-and-whisker' plot, see Figure 5.2).

BOXPLOT

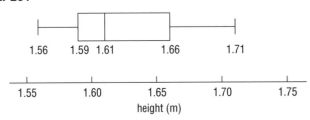

Figure 5.2 Boxplot drawn from the five-figure summary in Figure 5.1

The significant parts of the boxplot are shown in Figure 5.3.

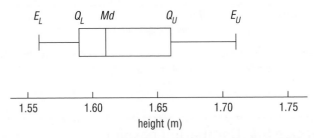

Figure 5.3 Boxplot showing where the five-figure summary values are placed

The central rectangle, which marks out the two quartiles, is called the 'box', while the two horizontal lines on either side are the 'whiskers'. Just by observing the size and balance of the box and the whiskered components we can gain a very quick and useful overall impression of how the batch of data is distributed. Also, drawing two boxplots one above the other can provide a powerful means of comparing two distributions. For example, have a look at Table 5.10, which compares wing lengths, in millimetres, of two very large samples of male and female meadow pipits. The corresponding boxplots are in Figure 5.4.

Table 5.10 Wing lengths, in millimetres, of meadow pipits

	Females	Males
Highest decile	82	85
Upper quartile	80	83
Median	77	81
Lower quartile	76	80
Lowest decile	74	78

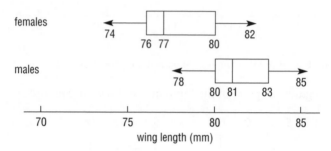

Figure 5.4 Decile boxplots of wing lengths, in millimetres, of meadow pipits
Source: adapted from Open University course 'Using Mathematics, MST 121', Chapter D4.

Exercise 5.6 Comparing boxplots

From the boxplots above, what general features can you pick out about wing lengths of the male and female meadow pipits?

[Comments below]

There are several clear patterns here, four of which are mentioned below.

▶ Notice that the 'whiskers' end with arrow heads, rather than short vertical lines. This is because these end points are 'deciles' (the lowest decile cuts off the bottom 10% and the highest decile cuts off the top 10% of the values) rather than the lower and upper extreme values. With very large batches (these data are based on a large batch size of birds) it really isn't very relevant to know what the extreme values are and so we tend to use the deciles instead.

- The wing lengths of the male birds are noticeably larger than those of the females. For example, the upper quartile of the female wing length coincides with the lower quartile of that of the males.

- Female wing lengths are slightly more widely spread than those of the males. Compare the lengths of the boxes between female and male and observe that the interval between the quartiles for female wing lengths is slightly wider. The same is true of the intervals between the arrow heads (the inter-decile range).

- Both boxplots are slightly skewed to the right – notice how both of the right-hand parts of each box are slightly longer than their equivalent left-hand parts. Incidentally, the direction by which skewness is described – skewed to the right or to the left – may seem counter-intuitive. It may help you to remember that the side which has the long tail is the direction to which we say it is skewed.

RANGE

The range is the simplest possible measure of spread and is the difference between the upper extreme value (E_U) and the lower extreme value (E_L). Returning to the height example, the tallest person is Rajvinder at 1.71 m and the shortest is Roberta at 1.56 m. Thus:

$$\text{Range} = E_U - E_L$$

so range = 1.71 m – 1.56 m = 15 cm.

One disadvantage of the range as a measure of spread is that it is strongly affected by an extreme or untypical value. For example, it only takes one extremely tall or extremely short person in the sample to have an enormous effect on the value of the range. This problem can be overcome by choosing to measure the range, not between the two extreme values, but between two other values on either side of the middle value. And what could be two more appropriate values to choose than the upper and lower quartiles!

INTER-QUARTILE RANGE (dq)

The inter-quartile range, sometimes known as the inter-quartile deviation, is, as you might expect, the difference between the two quartiles. In this example it is the difference between the heights of Sumita and Marcie.

> The inter-quartile range, $dq = Q_U - Q_L$

so inter-quartile range = 1.66 m – 1.59 m = 7 cm.

The inter-quartile range therefore contains the middle half of the batch values, which is, of course, the central box part of the boxplot.

A difficulty with any measure of spread which is tied to the units of the batch values is that it makes comparisons with the spread of other batches of data very misleading. For example, how might you compare the spread of a group of people's heights, measured in metres, with the spread of their weights, measured in kg? And would the spread of heights suddenly increase by a factor of 100 if the data values were converted from metres to centimetres? Even comparing the spreads of two batches measured in the same units can be misleading, as the next exercise shows.

Exercise 5.7 Comparing spreads

Using the data from Table 5.10, calculate the inter-quartile ranges for male and female wing lengths of the meadow pipits. What does this reveal about how widely spread their wing lengths are?

[Comments below]

dq *for female wing lengths*	*80 – 76 = 4 mm*
dq *for male wing lengths*	*83 – 80 = 3 mm*

We can see that the *dq* value for females, 4 mm, is slightly more than the *dq* value for males of 3 mm. So, on the face of it, female wing lengths seem to be more widely spread than those of males. However, this is a slightly misleading conclusion because we are not comparing like with like. Since male wing lengths are slightly more than females' we would expect any calculation based on male wing lengths to produce a larger result than a corresponding calculation based on female wing lengths. What we need to do is to *standardize* the inter-quartile range so that its value is independent of the actual batch values. This is done by expressing each *dq* value as a proportion of its corresponding median value, thus:

$$\text{Standardized inter-quartile range} = \frac{dq}{Md}$$

Standardized inter-quartile range for female wing lengths

$$= \frac{4}{77} = 0.052$$

Standardized inter-quartile range for male wing lengths

$$= \frac{3}{81} = 0.037$$

This standardized result suggests that, when expressed as a proportion of the median, the spread of female wing lengths is, in fact, a lot greater than that of males. Another example of standardizing the spread is provided later in the chapter when we discuss standard deviation.

MEAN DEVIATION

The mean deviation is a fairly commonsense measure of spread. It can be described as the 'average deviation of each batch value from the mean'. You will see how this measure is worked out in practice, but it will be introduced first with a deliberate mistake. As you work through the calculation, see if you can spot the error!

Here are the main stages involved in calculating the mean deviation. (Watch out for some silliness on the way.)

Table 5.11 shows how these instructions apply to the height data.

i Calculate the overall mean.

ii Subtract the mean from each value in the batch. This produces a set of deviations, *d* (where $d = X - \overline{X}$ for each value of *X*).

iii Find the mean of these deviations. The result is the 'mean deviation'.

Table 5.11 Incorrect calculation of the mean deviation

Name	Height (X) metres	Deviations (d)
Roberta	1.56	-0.062
Jenny	1.58	-0.042
Marcie	1.59	-0.032
Hilary	1.60	-0.022
Sue	1.61	-0.012
Alice	1.62	-0.002
Sumita	1.66	0.038
Fiona	1.67	0.048
Rajvinder	1.71	0.088
Sum	14.60	0
Mean	1.622	0

Stage (ii) (pointing to deviations -0.062, -0.042, -0.032, -0.022, -0.012)

Stage (i) (pointing to 1.71)

Stage (iii) rather silly here! (pointing to Mean 0)

As you can see, the calculation as it has been done here has produced an answer of zero for the mean deviation. Clearly, since there *is* a spread around the mean, this cannot be correct. So what has gone wrong?

The explanation lies with a basic property of the mean, namely that the sum of the positive deviations is exactly matched by the sum of the negative deviations. Therefore, the deviations *must* sum to zero – if they didn't, you would know that you had made a mistake! The problem is solved by treating all the deviations as positive. The mean deviation can now be defined more properly as the 'mean of the modulus of the deviations around $\overline{X}$' (where the term 'modulus' means 'the positive value of'). This is written in symbols as follows:

Mean deviation, MD, $= \dfrac{\Sigma|d|}{n}$

Said as 'sigma modulus *d* over *n*'. The modulus is written with two vertical lines, as shown here

Here, now, are the correct instructions for calculating the mean deviation.

i Calculate the overall mean.

ii Subtract the mean from each value in the batch. This produces a set of deviations, d (where $d = X - \overline{X}$ for each value of X). Take the modulus of these deviations, $|d|$.

iii Find the mean of these positive deviations – giving the 'mean deviation'.

Table 5.12 now shows how to calculate the mean deviation correctly!

Table 5.12 Mean deviation, MD = 0.0384 m, or 3.84 cm

Name	Height (X) metres	Modulus of the deviations (|d|)
Roberta	1.56	0.062
Jenny	1.58	0.042 — Stage (ii)
Marcie	1.59	0.032
Hilary	1.60	0.022
Sue	1.61	0.012
Alice	1.62	0.002
Sumita	1.66	0.038
Fiona	1.67	0.048
Rajvinder	1.71 — Stage (i)	0.088 — Stage (iii)
Sum	14.60	0.346
Mean	1.622	0.0384

VARIANCE (σ^2) AND STANDARD DEVIATION (σ)

The reason that the mean deviation (above) was introduced via a deliberate error, apart from being a desperate attempt to keep the reader awake, was in order to raise the question of how we should deal with the fact that the sum of the deviations around the mean is zero. The problem was solved for the mean deviation simply by making all the negative numbers positive. An alternative way of turning negative numbers into positive ones is to square them. This is the approach used when calculating the variance and the standard deviation. These two measures of spread are closely linked: the variance is the square of the standard deviation.

The symbol for variance is σ^2, while standard deviation is written as σ. The letter σ is the lower case form of the Greek letter 'sigma', not to be confused with the upper case version, Σ, which means 'the sum of'.

The method of calculating the variance is described below. As you read the description, glance back to how the mean deviation is calculated. Notice what is the same and what is different in the two explanations.

i Calculate the overall mean.

ii Subtract the mean from each value in the batch. This produces a set of deviations, d (where $d = X - \overline{X}$ for each value of X). Take the square of these deviations, d^2.

iii Find the mean of these squared deviations $\dfrac{\Sigma d^2}{n}$. The result is the 'variance'.

iv If you wish to find the standard deviation, take the square root of the variance.

Now, here are the formulas for these two measures of spread.

$$\text{Variance, } \sigma^2 = \frac{\Sigma d^2}{n}$$

$$\text{Standard deviation, } \sigma = \sqrt{\frac{\Sigma d^2}{n}}$$

Finally, Table 5.13 shows how to calculate the variance and standard deviation, using the height data.

Table 5.13 Calculation of the variance and standard deviation

Name	Height (X) metres	Deviations (d)	Squared deviations (d²)
Roberta	1.56	-0.062	0.003 872
Jenny	1.58	-0.042	0.001 783
Marcie	1.59	-0.032	0.001 038
Hilary	1.60	-0.022	0.000 494
Sue	1.61	-0.012	0.000 149

Stage (ii)

Name	Height (X) metres	Deviations (d)	Squared deviations (d²)
Alice	1.62	-0.002	0.000 005
Sumita	1.66	0.038	0.001 427
Fiona	1.67	0.048	0.002 283
Rajvinder	<u>1.71</u>	<u>0.088</u>	<u>0.007 705</u>
Sum	14.60	0	0.018 756
Mean	1.622	0	0.002 084
			0.045 650

Stage (i)

Stage (iii), variance

Stage (iv), standard deviation

So, variance = 0.002 084 m² and standard deviation = 0.045 650 m. If this were a practical example, these results would, of course, be rounded to an appropriate number of decimal places, depending on the circumstances of the question.

The procedure for calculating the variance and standard deviation is slightly more complicated when applied to data that have been organized into a frequency or percentage table. However, the method is exactly equivalent to the approach used when calculating a weighted arithmetic mean; namely, you have to remember to multiply each value (in this case, each squared deviation) by its corresponding weight, f, before finding the sum of the squared deviations. The formulas for finding a weighted variance and standard deviation are given below.

$$\text{Variance, } \sigma^2 = \frac{\sum fd^2}{\sum f}$$

Where $d = X - \bar{X}$ for each value of X

$$\text{Standard deviation, } \sigma = \sqrt{\frac{\sum fd^2}{\sum f}}$$

Remember that $\sum f = n$

There are other formulas for calculating the variance and standard deviation, which will produce the same results but make for a slightly easier calculation. For reasons of space these are not explained here, but increasingly these short-cut methods are becoming irrelevant as more and more users are performing such calculations on machines.

Finally, it is worth noting that of all the measures of spread that are available, the standard deviation is the one most widely used.

Nugget: two standard deviations

There is a tricky little complication about standard deviation that I just thought I'd slip in at the end of this chapter! There are actually not one but two standard deviations and which one you use depends on how you are planning to use it. The two scenarios follow.

If you have a set of numbers and you want to calculate their standard deviation because you want to know about their spread, then simply follow the instructions that I provided above. This calculation gives what is referred to as the *population standard deviation*, so named because the 'set of numbers' represents a complete set of figures and has not been taken from some wider population that you wish to know about.

Now consider a second very common scenario. Suppose that you wish to estimate the standard deviation of a particular population and to do so you take a sample from that population. Unfortunately, simply applying the previous standard deviation formula to the values in this sample will slightly underestimate the true population standard deviation (this is because the sample standard deviation is a biased estimator of the population standard deviation). A simple adjustment in the formula will correct for this bias and this adjusted formula for standard deviation is known as the *sample standard deviation*.

In textbooks, the population standard deviation is usually denoted by the Greek letter σ (said as 'sigma') whereas the sample standard deviation is often denoted by s.

Comments on exercises

▶ **Exercise 5.2**

a The mode is 'brown', with four occurrences.

b The mode is 7, with five occurrences.

c As is explained in the main text, calculating the mode with these data is meaningless.

▶ **Exercise 5.5**

The median for 'today' is two people per household, which is less than the 1961 figure.

Key ideas

This chapter dealt with a variety of statistical techniques for summarizing data.

▶ First, under the general theme of finding a typical or average value, we looked at the mean, the mode and the median.

▶ The second key question central to summarizing data was, 'How widely spread are the figures?', and this led to the following measures of spread – five-figure summary, boxplot, range, inter-quartile range, mean deviation, variance and standard deviation.

To find out more via a series of videos, please download our free app, Teach Yourself Library, *from the App Store or Google Play.*

6

Lies and statistics

In this chapter you will learn:

- ▶ *how to recognize misleading graphs*
- ▶ *how to know when people are cheating with percentages*
- ▶ *how averages can be used to mislead.*

It's not the size of your statistic, it's what you do with it that counts.

A variety of distortions, both deliberate and innocent, are commonly employed when numbers are used to bolster an argument. There is much scope in statistics for a little 'creativity', particularly when it comes to beguiling with a graph, being less than accurate with averages or even being perfidious with a percentage. The three sections of this chapter deal with these three potentially misleading areas of statistics – graphs, percentages and averages.

Misleading graphs

The first and most important thing to remember about drawing sensible graphs is that you need to be clear about the *purpose* for which the graph is to be drawn. The implication of this is that there is no single 'correct' graph for a particular data set – it all depends on what point you want the graph to make for your data. Having said that, there are a number of situations when a particular graph is clearly wrong for the type of data it has been used to represent and some examples are given here.

This section on graphs is based on four data sets, A, B, C and D, and a variety of possible graphical representations is offered for each. Some of the graphs suggested are quite sensible, while others are plain daft! Comments are given at the end of each collection of graphs but, before reading them, spend some time yourself looking at the suggested graphs and try to decide which are sensible and which are silly.

DATA SET A

Table 6.1 Data set A: times (seconds) for the egg and spoon race at Cookstown village fete

Names	Times
Anthony	26
Emma	18
Jaspal	19.6
Lisa	21
Meena	22

Names	Times
Navtej	27
Nicola	23
Sandeep	17
Tanya	23
Thomas	19

First a word about Data set A. The first column shows a set of names, so this is a set of *discrete* data. The second column gives running times and time is a *continuous* measure. What sort of graph you choose here will depend on whether you want the data to be organized principally on the basis of the names or the times. In general, discrete data are depicted using a bar chart or pie chart, whereas continuous data are best represented with a histogram.

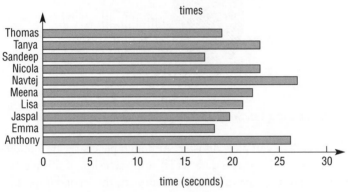

Graph A1 Horizontal bar chart showing the data in Table 6.1

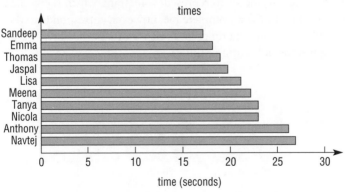

Graph A2 Ordered horizontal bar chart showing the data in Table 6.1

Graphs A1 and A2 show this data set represented in bar charts where the bars have been drawn horizontally rather than in the usual vertical orientation. Drawing bar charts this way round is a useful device when there is a large number of bars and there would be too much writing to put the labels onto the horizontal axis. Graph A2 is more helpful than A1 here, since it has placed the bars in rank order, showing clearly that Sandeep is the quickest and Navtej the slowest with an egg and spoon. However, both these graphs have the drawback that they don't show the distribution of data.

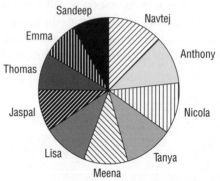

Graph A3 Pie chart showing the data in Table 6.1

Graph A3 is a complete disaster! Although pie charts are suitable for discrete data (as this data set is), and each slice of this pie does correspond to a separate 'discrete' category, the second important condition for using a pie chart has not been met here. This is that the complete pie needs to represent something which is complete and meaningful. This complete pie, however, corresponds to the sum of all the running times and this is not a particularly useful or interesting collection of things.

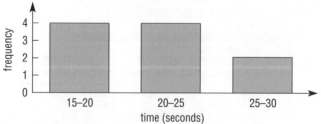

Graph A4 Block graph showing the data in Table 6.1

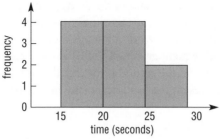
Graph A5 Histogram showing the data in Table 6.1

A different approach is taken in Graphs A4 and A5, where the focus of interest is overall patterns in the running times and the people's names have been ignored. These graphs have been produced as a result of grouping the times into 5-second intervals, counting the frequencies in each group and drawing them as a block graph. A weakness of Graph A4 is that it has been drawn with gaps between adjacent columns. As was explained in Chapter 4, this practice should only be used when depicting discrete data, and never for depicting continuous data such as time.

▶ Verdict

The most favoured graphs here are A2 and A5 but they each do a quite different job. A2 preserves the raw data and allows you to compare the relative times of each person in the race. A5 gives a picture of the data as a whole and shows the time intervals in which most or fewest people fall. Whichever of these graphs you might choose will depend on the purpose for which you wished to draw the graph in the first place.

DATA SET B

Table 6.2 Data set B: wind in January

Wind type	Days
Strong wind	10
Calm	5
Gale	7
Light breeze	9
Total	31

Data set B is *discrete* and shows the 31 days of a particular January categorized into four headings of windiness. Two features of this table of data are worth noting as they caused problems in some of the subsequent graphs. The first is the inclusion of the 'Total' category and the second is the fact that no attempt has been made to present the headings in any logical order.

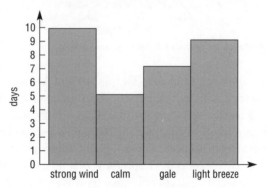

Graph B1 Block graph showing the data in Table 6.2

Graph B1 contains two problems. First, the sequence of wind categories is the same unhelpful one given in Table 6.2. A more logical ordering might be calm, light breeze, strong wind and gale. Second, the adjacent bars have been drawn so that they touch and this is not appropriate for discrete data.

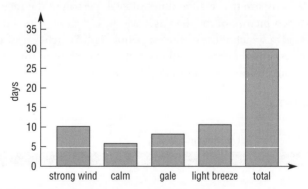

Graph B2 Bar chart showing the data in Table 6.2

Graph B2 has at least solved the problem of touching bars by drawing them with gaps but the sequence of the categories is still unsatisfactory.

Also the final 'Total' column which has been added is confusing as it implies that there are five categories when in fact the fifth is simply the sum of the first four columns.

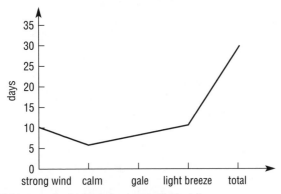

Graph B3 Line graph showing the data in Table 6.2

All of these drawbacks apply to Graph B3 but there is an additional twist. Joining the tops of the bars together here is meaningless for two reasons. First, such a technique is only sensible for continuous data and should therefore be reserved for situations where each point on the joining line or curve actually means something. For example, suppose you plotted a child's height on two successive Januaries. The line joining these points gives a reasonable estimate as to that child's height over the intervening months. This approach is not suitable for discrete data where it is hard to imagine what intervenes between the separate categories. This raises the second issue here, which is that the problem is made worse by the illogical order in which the categories have been presented. As a result, any pattern in the shape of the graph is totally spurious.

Most of these problems have been solved in Graphs B4 and B5. However, there is a problem with B4. Note how the vertical axis effectively begins at 4. This has a rather distorting effect on the columns. For example, the 'strong wind' column has been drawn to be six times as tall as the 'calm' column when in fact it only has twice the frequency (10 compared with 5).

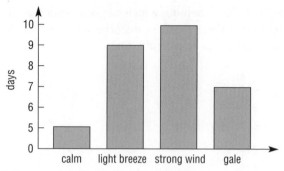

Graph B4 Bar chart showing the data in Table 6.2

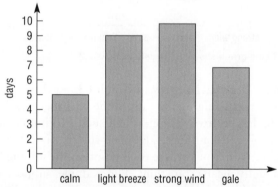

Graph B5 Bar chart showing the data in Table 6.2

While it is not 'illegal' to draw the vertical axis so that it starts with a non-zero value, the usual practice is to show a 'break' in the axis as a way of alerting the reader to the potentially misleading interpretation. This is shown below.

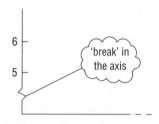

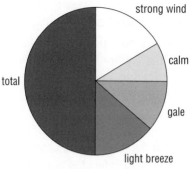

Graph B6 Pie chart showing the data in Table 6.2

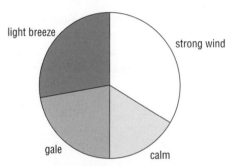

Graph B7 Pie chart showing the data in Table 6.2

And finally to the two pie charts. The good news is that this data set is highly suitable for a pie chart because it satisfies the two basic conditions, namely that the data are discrete (yes, they are) and that, taken as a whole, the complete pie means something. This second condition certainly holds for Graph B7 since the whole pie represents the 31 days of January. However, Graph B6 is a bit daft since the 31 days have been included twice – once in the four small slices and then again in the total.

▶ Verdict

Only two of these graphs appear to be satisfactory, the bar chart B5 and the pie chart B7.

Nugget: stretching and squeezing

Here is quite an entertaining computer-based exercise that will illustrate just how easily a graph's shape can be distorted so that it can be made to convey a very different impression, simply by adjusting the scales on the axes.

First, find or make a line graph on a computer (you could search for one on a search engine such as Google or create your own using a spreadsheet or database). Next, paste this graph into a word-processing document. Select the graph by clicking on it. You can now click-and-drag the image using any of the four corners of the image. This has the effect of altering the size of the graph, but not its shape, thereby preserving the slope of the graph. Now do the same thing but this time using one of the four side handles of the image. This time the size alters in one dimension only, so the shape will change. You may be surprised at the corresponding effect on the slope of the graph.

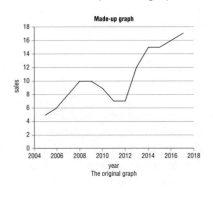

The original graph

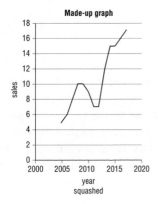

squashed

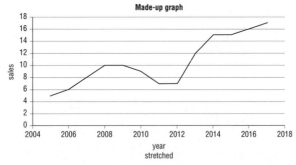

stretched

DATA SET C

Table 6.3 Data set C: Queen cake ingredients

Ingredients	Measurements
Margarine	6 oz
Self-raising flour	1 lb
Caster sugar	6 oz
Sultanas	2 oz
Vanilla essence	¼ tsp
Eggs	3

One characteristic of recipe ingredients is that they tend to be measured in different units. For example, weights may be in ounces, pounds, grams and so on. Some ingredients aren't weighed at all – eggs, for example – while sometimes non-standard measures are used – teaspoons, cups and so on. As a result, the *numbers* used to measure the different ingredients *are not comparable with each other*. Since the whole point of graphs is to be able to compare numbers visually, this gets Data set C off to a bad start. Indeed, it is hard to imagine why one might ever wish to graph these data!

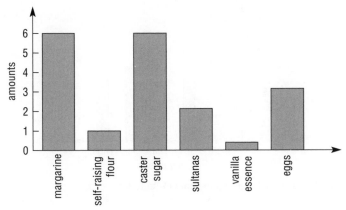

Graph C1 Bar chart showing the data in Table 6.3

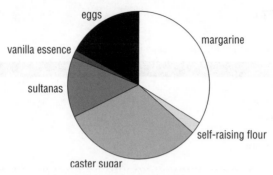

Graph C2 Pie chart showing the data in Table 6.3

Because of this problem of differing units, it is a meaningless exercise to compare the heights of the columns in the bar chart (Graph C1) and assess the relative sizes of the slices of the pie chart (Graph C2). For example, according to the graphs, you can see that this cake appears to be seriously under-supplied with flour. This is because the flour is measured in pounds, whereas the sugar and margarine are given in ounces. Similarly, there seem to be half as many eggs (three) as there are margarines (six)! Hmmm! What is going on here?

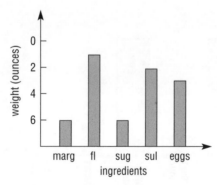

Graph C3 Bar chart showing the data in Table 6.3

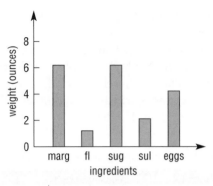

Graph C4 Bar chart showing the data in Table 6.3

Graphs C3 and C4 compound all of the above problems with some additional foolishness along the vertical axis. C3 has contrived to have the vertical scale the wrong way round, which has resulted in the largest data values having the shortest bars, and vice versa. The vertical scale in graph C4 has been drawn with unequal intervals which means that you can't trust the height of each bar to give an honest impression of the value it is meant to represent.

▶ **Verdict**

Forget it!

Confusing percentages

> .... If the standard of driving on motorways justified it, then I think a case for raising the speed limit would exist. As it is, surveys have shown that over 85% of drivers consider themselves to be above average! In reality, the reverse is more likely to be true
>
> From a letter to *Motorcycle Rider*, Number 13

As was indicated in Chapter 5, percentages are extremely useful for making fair comparisons between two numbers or two sets of numbers. For example, Table 6.4 shows the figures for church membership in the United Kingdom over a 15-year period coming up to the end of the last millennium. The data have been presented in three categories: Trinitarian churches (including Anglican, Presbyterian, Methodist, Baptist and Roman Catholic), non-Trinitarian churches (including Mormons, Jehovah's Witnesses and Spiritualists) and other religions (including Muslims, Sikhs, Hindus and Jews).

Table 6.4 Adult church membership in the United Kingdom (millions)

	Base year	15 years later
Trinitarian churches	8.06	6.77
Non-Trinitarian churches	0.33	0.46
Other religions	0.81	1.86

Source: adapted from Social Trends 22.

Because these figures differ so widely, it is hard to make direct comparisons between the different religions and say how they have declined or grown. In this sort of situation, calculating the percentage changes is a useful way of comparing the relative changes for each religious group. Thus, relative change of membership for:

$$\text{Trinitarian churches} = \frac{6.77 - 8.06}{8.06} = \frac{-1.29}{8.06}$$
$$= -0.160 \text{ or } -16.0\%$$

$$\text{non-Trinitarian churches} = \frac{0.46 - 0.33}{0.33} = \frac{0.13}{0.33}$$
$$= 0.394 \text{ or } 39.4\%$$

$$\text{other religion} = \frac{1.86 - 0.81}{0.81} = \frac{1.05}{0.81}$$
$$= 1.296 \text{ or } 129.6\%$$

Using these percentage changes we can now compare the rates of growth or decline directly over the 15-year period. Roughly speaking, the Trinitarian churches have lost around one in six of their flock while the non-Trinitarian churches have gained about two members for every five that they had at the start of the period. The other religions have grown more dramatically, with their numbers having more than doubled.

'A welcome slowdown' was how, in the 1990s, John Patten, then the UK Home Office minister, described the 16% rise in crime figures one year, which was less than in previous years. This tactic is very much the stock in trade of politicians when asked to respond to an unpalatable percentage, namely to ignore the size of the change and concentrate only on whether the rate of change is going up or going down. Note that this 'slowdown' still represents a 1 in 6 increase over the previous year, so reported crime rates were still rising rapidly. To claim that the *rate* of increase is not quite as fast as it was before is a little disingenuous – a bit like saying that we lost the match 10–0 but at least we improved on the 12–0 drubbing we sustained the week before! The same gambit is used over the rate of inflation. When a politician claims that the rate of inflation has fallen, many people mistakenly believe that this implies that *prices* must therefore have fallen. On the contrary, any positive rate of inflation, whether it happens to be rising or falling, reflects rising prices. For the rate of inflation to fall from say, 8% to 6% simply means that prices are continuing to rise but at a slightly slower rate than before.

Nugget: advertiser-speak

On a TV advertisement I viewed recently, a container is seen being filled with washing gel. The voice-over commenting on this particular gel states that: 'every drop has 65% more concentrated freshness'.

This claim raises three questions for me:

▶ What exactly is 'concentrated freshness' and how would I measure it?

▶ 65% more than *what*?

▶ Why do advertisers continue to patronize us with this sort of meaningless nonsense?

Staying in the world of politics, here is an example where a percentage was not used but perhaps should have been. The theme is poverty.

What does it mean to be 'in poverty'? Clearly, one person's notion of poverty will not be the same as another's. To some extent, our view of poverty is relative to the level of wealth around us – a poor family in the USA could be seen as being very well-off in, say, Ethiopia during a major famine.

Governments and other organizations such as the Low Pay Unit are interested in analysing the patterns in poverty so that the groups in greatest need can be clearly identified. An essential element in any such investigation is to have agreement between government and pressure groups on some unambiguous criterion for measuring what we mean by 'low pay'. An agreed measure of 'low pay' in the UK is earnings which are less than 50% of average (median) income. The figures in Table 6.5 indicate the relative numbers of children in poverty in the UK, by status of household.

Table 6.5 Typical numbers of dependent children in households below 50% of average income, analysed by family type

Status of household	Number of children (000)
Parent in full-time work	1025
Lone parent	938
Unemployed parent	763
Pensioner	320

Source: estimated from national UK data.

Exercise 6.1 What's that as a percentage?

Convert the figures from Table 6.5 to percentages. What do they reveal about which of the four categories of household contains the greatest number of children in poverty?

[Comments below]

The total number of children here, measured in thousands, is

$$1025 + 938 + 763 + 320 = 3046$$

The percentage figures are calculated as follows:

Status of household	% Children
Parent in full-time work	$\frac{1025}{3046} \times 100 = 34$
Lone parent	$\frac{938}{3046} \times 100 = 31$
Unemployed parent	$\frac{763}{3046} \times 100 = 25$
Pensioner	$\frac{320}{3046} \times 100 = 11$

Representing these percentages on a bar chart produces the graph in Figure 6.1.

The message of this figure seems to be that the largest number of children in poverty is associated with households where there is a parent in full-time work. However, although this conclusion is not wrong, the percentages which we have calculated do not tell the whole story and are therefore rather misleading.

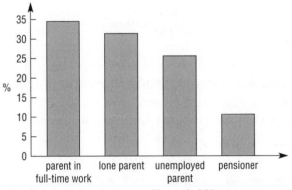

Figure 6.1 Bar chart showing types of household in greatest poverty

What is missing from this analysis is the total numbers of children in each of these four categories. These data have been added in Table 6.6.

Table 6.6 Number of children in households earning below 50% of average income, analysed by family type

Status of household	Number of children in poverty (000)	Total number of children (000)
Parent in full-time work	1025	9330
Lone parent	938	1250
Unemployed parent	763	930
Pensioner	320	470
Total		11 980

Source: estimated from national UK data.

As can be seen from the last column in this table, there are many more children from households in the category 'parent in full-time work' than from any other. It is therefore not surprising that this category of household produces the largest number of children in poverty. What is a more useful calculation in this context is the percentage of each separate group of households in poverty. For example,

Percentage of children in poverty from lone-parent households

$$= \frac{938}{1250} \times 100 = 75\%$$

Exercise 6.2 Recalculating the percentages

Calculate the other three percentages in this way and draw the results as a bar chart. How do you interpret these figures?

[Comments below]

When the percentages are calculated in this new way, based on the total number of children in each type of household (see Table 6.7 and Figure 6.2), it is clear that the majority of children from lone parent, unemployed parent and pensioner headed households are in poverty while only about one in ten children from full-time worker headed households is similarly economically disadvantaged.

Table 6.7 Percentage of children in poverty based on size of each household status category

Status of household	% Children
Parent in full-time work	$\frac{1025}{9330} \times 100 = 11$
Lone parent	$\frac{938}{1250} \times 100 = 75$
Unemployed parent	$\frac{763}{930} \times 100 = 82$
Pensioner	$\frac{320}{470} \times 100 = 68$

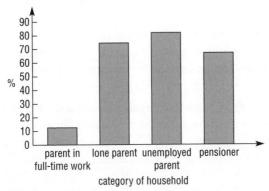

Figure 6.2 Bar chart showing percentages in Table 6.7

Nugget: percentage poser

Here is a little conundrum about percentages. Let's take any starting number (for convenience, I'll choose 100), increase it by, say, 20%, then reduce the answer by *the same* percentage (in this case, 20%), then logic would suggest that you get back to the original starting number ... well do you?

Let's try it and see. Try following through the numbers in the table below. And use your calculator just to check my arithmetic.

Step	Value
1. Take any starting number	100
2. Increase it by 20%	20% of 100 = 20 100 + 20 = 120
3. Decrease the answer by 20%	20% of 120 = 24 120 – 24 = 96

Now try this same calculation for different starting numbers and different percentages (always remembering to use the same percentage increases and decreases each time).

What exactly is going on here...?

Inappropriate averages

Overheard in a bar in Glasgow:

When a Scot emigrates to England it raises the average level of intelligence in both countries.

We shall start with a situation where calculating an average might have been helpful but it wasn't actually used. The following cutting comes from an article on the British car industry.

Ford's job losses will reduce its manual work force to 14 000. This year it plans to make 1.6 million cars. But Nissan's factory in Sunderland will be producing 500 000 cars a year by next year with a workforce of just 6200. Toyota's annual target is 200 000 cars with 3800 workers.

Exercise 6.3 Interpreting a jumble of figures

What do these figures appear to say about the relative productivity of Ford, Nissan and Toyota?

[Comments below]

When contained within a piece of text like this, the six numbers mentioned are rather difficult to interpret. It may clarify things to present the numerical information as shown in Table 6.8.

Table 6.8 A neater summary of the data

	Cars	Workforce
Ford	1 600 000	14 000
Nissan	50 000	6200
Toyota	200 000	3800

The whole thrust of the article accompanying this table was that, with its very high levels of production, the Ford car-making process was also more efficient in its use of workers than that of the two Japanese firms. If this were true, a simple average would make this point rather more clearly, as illustrated in Table 6.9. As you can see from the final column of the table, this would indeed seem to be the case as the average number of cars per worker for Ford is considerably higher than for the other two companies. However, the comparison is not entirely valid for a different reason not mentioned in the article, namely that the Japanese companies also manufacture engines for export and these were not included in the calculation.

Table 6.9 Including the averages

	Cars	Workforce	Average number of cars per worker
Ford	1 600 000	14 000	$\frac{1\ 600\ 000}{14\ 000} = 114.3$
Nissan	500 000	6200	$\frac{500\ 000}{6200} = 80.6$
Toyota	200 000	3300	$\frac{200\ 000}{3300} = 60.6$

For another example of the way that averages can affect the way we make sense of information, let us return to the late 1980s, when Margaret Thatcher, the then prime minister of the UK, stated that:

> *Everyone in the nation has benefited from the increased prosperity – everyone.*

Her remarks point to a common error with averages, which is to believe that they provide information about the *spread* of values, or in this case about the distribution of income and wealth. In fact, averages provide no information about the spread of values within a data set. Indeed, as the following simple example shows, it is possible for the average earnings to rise while the majority of people are actually worse off.

Imagine a small firm with five employees, all of whom were earning, say, £100 per week. The following year, the salaries were regraded. Four of the employees had their salaries reduced to £90, while the lucky fifth got a rise to £180.

Exercise 6.4 Far better or far worse?

What effect did this have on the average earnings of the five employees?

[Comments below]

Average earnings before the regrading

$$= \frac{100 + 100 + 100 + 100 + 100}{5} = £100$$

Average earnings after the regrading

$$= \frac{90 + 90 + 90 + 90 + 180}{5} = £108$$

So, the average earnings have increased (by 8%) but actually four out of the five people are less well off than before.

This is clearly a highly artificial example but in principle it describes what happened in the UK during the 1980s under Mrs Thatcher's government. Although average real earnings (i.e. earnings taking account of the effects of inflation) rose during this decade by something of the order of 30%, the widening inequalities resulted in many more people falling into debt and poverty. For example, have a look at the figures in Table 6.10.

Table 6.10 The distribution of the gross weekly earnings (rounded to the nearest £1) of men and women in April 1980 and April 1990, covering the 'Thatcher era' (1979–1990)

	April 1980		April 1990	
	Men	**Women**	**Men**	**Women**
Highest decile	183	116	467	317
Upper quartile	143	91	347	245
Median	113	72	258	177
Lower quartile	90	58	193	136
Lowest decile	74	49	150	111

How, then, can we compare the spread of earnings for these four columns of data? The simplest measure of spread to use here is the inter-quartile range. For example, taking the earnings column for women in April 1980:

$$\text{Inter-quartile range} = £91 - £58 = £33$$

Exercise 6.5 Calculating inter-quartile ranges

Calculate the inter-quartile ranges for the other three columns of data. Use your answers to compare the spreads of earnings for men and women over the decade between 1980 and 1990.

[Comments at the end of the chapter]

Clearly the inter-quartile ranges for 1990 (£154 for men and £109 for women) are much wider than for 1980 (£53 for men and £33 for women). However, since the numbers in the 1990 columns are larger than in the 1980 columns, it isn't very surprising that the inter-quartile ranges for 1990 should work out to be more than for 1980. You may remember that this issue cropped up in the previous chapter. There it was suggested that the only way a fair comparison can be made between two measures of spread is if they are *standardized*. The usual way to standardize the inter-quartile range is to divide it by the median. For example, the calculation for 1980 women is as follows:

$$\text{Standardized inter-quartile range for 1980 women} = \frac{33}{72} = 0.46$$

Exercise 6.6 Standardizing the inter-quartile range

Calculate the standardized inter-quartile range for the other three columns of data. Use your answers to compare the spreads of earnings for men and women over the decade between 1980 and 1990.

[Comments below]

On the basis of the standardized inter-quartile ranges calculated in Exercise 6.6, there is now some objective evidence of how earnings inequalities widened over the 1980s. The standardized inter-quartile ranges for both men and women in 1980 were just under 0.5 and by 1990 they had risen to around 0.6.

Comments on exercises

▶ **Exercise 6.5**

For 1980 men, inter-quartile range = £143 – £90 = £53

For 1990 men, inter-quartile range = £347 – £193 = £154

For 1990 women, inter-quartile range = £245 – £136 = £109

▶ **Exercise 6.6**

For 1980 men, standardized inter-quartile range $\dfrac{53}{113}$ = 0.47

For 1990 men, standardized inter-quartile range $\dfrac{154}{258}$ = 0.60

For 1990 women, standardized inter-quartile range $\dfrac{109}{177}$ = 0.62

Key ideas

This chapter provided some examples of ways in which statistics can mislead. It covered three key areas where statistical jiggery-pokery tends to flourish:

1 graphs

2 percentages

3 averages.

The chapter ended with a longer case study where a suitable measure of spread, rather than a simple average, was needed in order to come to a sensible conclusion.

To find out more via a series of videos, please download our free app, Teach Yourself Library, *from the App Store or Google Play.*

7

Choosing a sample

In this chapter you will learn:

▶ *the difference between a population, a sampling frame and a sample*

▶ *how to carry out random sampling*

▶ *how to generate and use random numbers when sampling*

▶ *the distinction between systematic and random sampling*

▶ *how errors can affect sampling.*

Keith Parchment was arrested in April 1986 on suspicion of robbery. A major dispute at his trial concerned whether he had made a true confession or whether, as he claimed, it had been falsified by the police. In court, the detectives stated that Mr Parchment's original confession had taken them 31 minutes to write down. This worked out at a rate of 174 characters per minute. In order to test this out, his statement was written out again by one of the investigating team. This experiment revealed that rewriting his statement took 44 minutes to complete. So, was the version provided by the police a forgery?

The key question here was just how likely it might be that someone could (or would choose to) write legibly without a break at a rate of 174 characters per minute for 31 minutes. On the basis of the one person tested, this does seem unlikely. However, perhaps the person tested was a particularly slow writer. Simply choosing a sample of one person does not seem sufficient to come to a conclusion with any degree of confidence. Taking the population as a whole, how likely is it that someone chosen at random would have a writing speed of 174 characters per minute? We could always test the entire adult population of the UK, but this would prove far too costly and time-consuming.

In the event, a sample was taken of writing speeds among 17 staff at a forensic science laboratory. The participants were asked to try to ensure that their writing was as legible as that on the statement. The results of this experiment produced writing speeds which ranged from a minimum of 124 characters a minute to a maximum of 158 characters a minute. On the basis of this larger sample, Keith Parchment now had much firmer evidence that his evidence had indeed been falsified by the police and this was a key factor in his subsequent acquittal.

This example illustrates some important features of sampling which will be highlighted in this chapter. In general, we choose a sample in order to measure some property of the wider population from which it was taken. This may be in order to discover certain information about a single population (the extent of childhood illness, how many cigarettes people smoke and so on), or the sampling may be part of an investigation into whether two or more populations are measurably different (are there regional differences in childhood illness, do boys smoke more than girls and so on). A 'good' sample is one which fairly represents the population from which it was taken.

Nugget: sampling soup

A remark that you sometimes hear when the results of some survey are being discussed is, 'Well, nobody asked *me*!' The whole point of surveying opinions by taking a sample is that, provided the sample is carefully selected so as to be representative of the entire population, it should provide an accurate overview without having to go through the cost and effort of asking everyone's opinion. An analogy that students find useful here is that of a cook making a large pot of soup. In order to check its flavour, she will first give the pot a good stir (to ensure that it is thoroughly and evenly mixed) and then sample a (representative) spoonful. Doing this avoids the need for her to consume the entire potful in order to make a sensible judgement about its flavour!

This chapter looks at some of the key issues involved in choosing a 'good' sample, and starts with a definition of what we mean by the term 'sampling frame'.

The sampling frame

When talking about sampling in statistics, it is useful to separate out three related but distinct levels of terminology. Starting at the top level first, these are known as the 'population', the 'sampling frame' and the 'sample'. Figure 7.1 below suggests that the sampling frame is (often) a subset of the population and the sample, in turn, is a subset of the sampling frame.

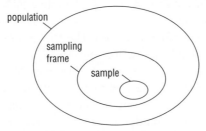

Figure 7.1 Population, sampling frame and sample

▶ **Population** – in statistics, this term 'population' isn't restricted in meaning to refer just to a population of animals or humans but is used to describe any large group of things that we are trying to measure. This might be a precise measure of the length of each item of a particular manufactured component, or the weight of each bag of crisps coming off a production line, or the life in hours of a particular brand of light bulbs. In general, manufacturers need to monitor their production processes to ensure that their grommets or their bags of crisps, etc. are continuing to meet the required standard. Also, governments expect to collect detailed information about many aspects of the human populations for whom they provide services and from whom they collect money. It would be too expensive and time-consuming to measure each grommet or weigh each bag of crisps or interview each person in the population, so they all usually resort to some sort of sampling technique. The size of the population, i.e. the number of items in the population, is usually denoted by an upper case N.

▶ **Sampling frame** – this is a list of the items from which the sample is to be chosen. Often it is the case that the sample is taken directly from the population, in which case the sampling frame would simply be a complete list cataloguing each item in the population. Sometimes, however, the population is simply too large and complicated for each element in it to be itemized accurately. For example, if the government wishes to estimate the national acreage of land growing oilseed rape, a suitable sampling frame would be a list of farms in the country. However, it is a major undertaking to list all the farms in the UK and it is likely that a number of smallholdings will be omitted.

▶ **Sample** – this, in turn, is the representative subset of the sampling frame which is chosen as fairly as possible to represent the entire population. The term 'sample size', usually denoted by a lower case letter 'n', refers to the number of items in the sample. Where more than one sample is taken, it is worth stressing that 'n' is the number of items selected in a particular sample and not the number of samples taken. The (rare) situation where the entire population is 'sampled', i.e. where $n = N$, is known as a 'census'.

The tale of Keith Parchment illustrates two very important features of sampling. First, if a sample is too small it may not provide a very reliable or representative cross-section of the population as a whole and indeed may give a wrong impression of it. Second, there is the other side to this coin, namely that large samples, although they may be highly representative and accurately reflect the population, are usually also very expensive and time-consuming to carry out. In practical terms, therefore, sampling is always a compromise between accuracy on the one hand and cost on the other.

Random sampling

There are several techniques for choosing a representative sample, of which probably the best known is *random sampling*. Its basic definition is as follows.

> A random sample is one in which every item in the population is equally likely to be chosen.

With random sampling, there also needs to be some sort of random-generating process used to select each item. Suppose, for example, we wished to select a sample of size 50 (sample size, $n = 50$) from a population consisting of a sampling frame of size 1000 (i.e. $N = 1000$). The usual approach would be to allocate a number to each item in the sampling frame (from 1 to 1000 or perhaps from 0 to 999) and then find some way of randomly generating 50 numbers from within this range. The population items which had been allocated these numbers then become the 50 items which form our sample.

There are two common methods of random sampling – sampling with replacement and without replacement. Replacement means simply putting an item back into the sampling frame after it has been selected. Although replacement is sometimes thought of as being a more correct form of random sampling, in most practical sampling situations items are not replaced after they have been selected.

A useful mental image for a method of generating random numbers is a bingo 'random number generator' (RNG). Typically this is a large Perspex box containing 90 ping-pong balls numbered 1 to 90, which are tossed around by jets of air. The balls are released through a chute and displayed for all to see. The bingo caller then reads out the numbers in the time-honoured fashion – 'legs-eleven', 'blind-forty', 'all the sixes, clickety click, sixty-six', etc. This continues until a punter's card has been filled and the prize is awarded. The balls are then put back into the RNG and the process begins again with a new game.

Exercise 7.1 Sampling with and without replacement

a As described above, is the bingo RNG an example of sampling with replacement or without replacement?

b Why do you think the bingo RNG is designed in Perspex?

[Comments below]

Once a bingo number has been selected, it is no longer available for reselection in the same game, so it is an example of sampling without replacement.

There may be several reasons why the box is made from Perspex. One may be that it is an interesting looking device which brings excitement and a sense of theatre to the game. But another important consideration is that the process is seen to be fair by all concerned. This is one reason why the ping-pong balls are usually set out on a rack for the punters to see once the balls have been selected. Clearly selections have the effect of producing winners and losers – of money, power, prestige and so on – and throughout history such events have been traditionally associated with cheating. So, being fair and being seen to be fair in the selection process are both essential requirements of random sampling.

There have been many random generators over the centuries which have served a rather different purpose than to select a representative sample. For thousands of years, humankind has tried to divine the future by interpreting the outcomes of chance events. Of course, those who believe in this notion of 'chance divination' may describe the process rather differently. According to Marshall Cavendish in the book *Paths to Prediction* (Marshall Cavendish Books Ltd, 1991), it is a process in which we 'give Fate an opportunity to make its intentions known by using the operations of Chance'. The view of life implicit in this quotation appears to be one in which our Fate has been pre-ordained and we can tap its knowledge by means of any number of random generators – cards, tea leaves, bumps on the head, coins, dice and so on. For example, *I Ching*, or *The Book of Changes*, in Chinese, is an ancient philosophy which is founded on the centrality of Fate, and indeed Confucius turned to it for guidance and information. The randomization process which it involves can be carried out by using either three brass coins or a bunch of yarrow sticks. In the case of the coins, the 'heads' is given a value of 3 and the 'tails' has a value of 2. The three coins are tossed, giving a combined score of 6, 7, 8 or 9. This is carried out six times in all, thereby generating a six-digit number, say, 687 986. *The Book of Changes* matches this number up to a fragment of poetry which, hopefully, reveals its mysteries in the context of the question you have asked.

Clearly there is more to doctrines like *I Ching* than the mere one-off prediction of a future event. At the heart of the divinations lies a set of 'truths' and teachings which can provide inspiration and guidance for the whole of a person's life and for their entire community.

However, the other aspect of such predictions is the underlying belief that the chance events in question do not produce arbitrary or random outcomes but that they are somehow controlled and influenced by some unspecified supernatural force. This runs exactly counter to the statistician's view, which is that each outcome of a truly random event is indeed arbitrary, and the only underlying pattern is the obvious statistical one of equal likelihood, i.e. that, in the long run, you would expect each number to come up roughly the same number of times (not *exactly* the same number of times because you would expect some degree of variation).

Let us now return to random sampling with a simple example. Suppose that you have bought a ticket for a concert and find you are unable to go. In an unaccustomed burst of generosity, you decide to give the ticket to one of three close friends. But all are equally deserving and you owe none of them a special favour. How do you choose which friend should receive the ticket?

Clearly this 'typical everyday' scenario is leading irrevocably to some sort of random selection procedure! One possibility might be to use the method of drawing straws and select them, in turn, on behalf of each friend. Whichever friend matched up with, say, the longest straw, would get the ticket. Another approach might be to roll a die. If the die shows a 1 or a 2, then friend A gets the ticket, if it shows a 3 or 4, then friend B gets it, and so on. But whichever approach you choose, two basic elements must be present, namely:

a each friend must be matched up to a particular object (such as a straw) or number or set of numbers (as in the case of the die);

b there needs to be a fair mechanism for selecting the object or number such that each has an equal chance of selection.

Exercise 7.2 Some folks have all the luck

Here are two alternative methods of selection. Are they fair and, if not, then why not?

Method 1	Spin a coin between two people, A and B. If it shows 'heads', then A loses, otherwise B loses. Now spin a coin between the winner and person C. Whoever wins this spin gets the ticket.
Method 2	Toss two coins. If they show double-heads, A wins. If they show double-tails, B wins. If they show one head and one tail, then person C wins.

[Comments below]

Both of these methods fail on the grounds that the odds of winning are not equally shared among the three people. The unfairness of Method 1 lies in the fact that person C has got a 'bye' into the second round and has only one hurdle to overcome in order to win the ticket, whereas A and B both have two hurdles to overcome. C is actually twice as likely to win as either A or B. Method 2 also has a bias in favour of person C, but for a different reason. The explanation may be clearer if the four possible outcomes are written out. These are HH, HT, TH and TT. Each of these four outcomes is equally likely, so there is a 1 in 4 chance of any one of them occurring. However, note that A wins only when HH appears and B wins only when TT appears, but C has two chances to win – C will win with either HT or TH. Thus, once again C is twice as likely to win the ticket as either A or B.

Nugget: equally likely

The basic ideas of probability are often (but not always) based on events with equally likely outcomes. Coins and dice provide a useful way to understanding these ideas. For example, events such as the sex of an unborn child, with just two equally likely outcomes, can be modelled by tossing a fair coin. Where there are six (roughly) equally likely outcomes, you could try rolling a die. For anything else, either use random number tables or the random command on a computer or calculator. The main reason that these everyday objects are used for this purpose is to do with their near-perfect symmetry, which ensures equal likelihood. A second advantage is that, although two-headed coins and loaded dice do exist, it is hard for someone to bend them to their will, however much they mumble incantations or blow on them beforehand. Fortunately, when it comes to chance events, Lady Luck tends to make her own choices!

Generating random numbers

Coins and dice are quite good random number generators in that their symmetrical shapes ensure fairly random sequences. However, for selecting samples of, say, 15 or 20 or more from larger populations, the whole thing becomes a bit of a chore. You may be thinking that it would all be much easier if the throws of the die had already been done for you and the scores were set out in a table. In fact, this is exactly what a *random number table* provides, although the numbers are randomized in the range 0 to 9, rather than 1 to 6.

Figure 7.2 shows an extract from a random number table of the sort that can be found at the back of many statistics textbooks. Such tables hardly make for gripping reading, but some points are worth noting. First, you can see how the numbers have been grouped in fives. This is simply to make it easier to read them. Also, unlike most sets of tables, there is no natural starting point – you can take any randomly chosen point in the table as the starting point and then simply read along the line from there, taking the numbers one at a time or two at a time, etc., whichever is appropriate for your needs.

00	81668	75363	62126	65806	71928	30458	17405	39056	52083	20028
01	09065	19470	15770	43347	41754	63327	09071	56236	63510	45541
02	32519	12965	30543	88542	18830	31744	00980	43126	32154	32796
03	98740	98054	30195	09891	18453	79464	01156	95522	06884	55073
04	85022	58736	12138	35146	62085	36170	25433	80787	96496	40579
05	17778	03840	21636	56269	08149	19001	67367	13138	02400	89515
06	92409	79891	60979	67158	07958	72053	36272	78804	42477	40338
07	49487	52802	62058	87822	14704	18519	17889	45869	36752	54958

etc.

Figure 7.2   Random number tables (an extract)

An alternative to using a random number table is to generate random numbers from a computer or calculator. Many scientific calculators possess a key marked something like 'Rand' or 'Ran#' which, typically, will produce a random decimal number in the range 0 to 1. Alternatively, it may be possible to use this facility to generate whole numbers randomly within a specified range.

Using random numbers to select a sample

Let us now look at how random number tables might be used in choosing a sample.

Here's a simple example of sampling in action, based on using the random number table in Figure 7.2.

In order to get student feedback on a statistics course that he had just taught, Alan decided to sample the opinions of 10 of his 43 students. Alan wanted to avoid choosing a biased sample, so decided to sample the ten students randomly. He went through the following five steps.

Step	Outcome
1. List the students' names in alphabetical order, numbering them 1 to 43. Note: in real life you are unlikely to hit upon 43 consecutively different names.	01 Alex 02 Anna 03 Broady 04 Dave etc.
2. Roll a die to select which row of random numbers to start with from Figure 7.2.	Alan rolled a '4'.
3. Read across line 04 of the table of random numbers, choosing the digits two at a time.	85, 02, 25, 87, 36, 12, 13, 83, ...
4. Make a record of the first ten selections, disregarding numbers greater than 43 and ignoring any number that has already been selected.	85, 02, 25, 87, 36, 12, 13, 83, 51, 46, 62, 08, 53, 61, 70, 25, 43, 38, 07, 87, 96, 49, 64, 05 This 25 is ignored, because 25 has already been selected.
5. List the ten student names corresponding to this selection of numbers. These students comprise the random sample.	02. Anna 13. Leo 05. Dee 25. Maddy 07. Erin 36. Pete 08. Fiona 38. Ravi 12. Hattie 43. Whitney

Sampling variation

Alan noticed that only three of the 10 students in his sample were male (Leo, Pete and Ravi), despite the fact that the 43 students taking the course were fairly evenly split male/female. You might be thinking that this represents a bias in the selection process but there is no systematic bias when the sample is selected randomly. A feature of any random selection process is that there is always likely to be a considerable degree of variation between one sample and the next. To illustrate this point, try tossing a coin ten times and make a note of the result. Although for each toss, getting either 'heads' or 'tails' are equally likely, there is no guarantee but you will get exactly 5 heads and 5 tails. In fact there is only about one in four chance of getting this exact even split, so getting outcomes such as 6 heads and 4 tails or 3 heads and 7 tails should not be a great surprise. Figure 7.3 shows the expected number of occurrences for each outcome for such an experiment, if a fair coin was tossed 1000 times. As the bar chart shows, only on roughly 250 occasions might you expect to get an even 5/5 split between heads and tails. On the other 750 occasions (three quarters of the time) you could expect some other outcome.

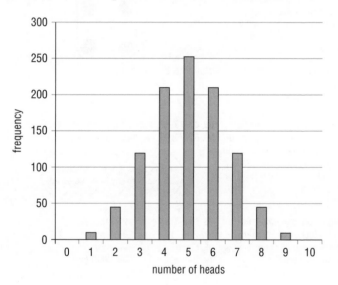

Figure 7.3 Expected outcomes for 1000 tosses of a fair coin

Sampling variation refers to the reality of what happens when samples are selected randomly from a wider population, namely that the samples will probably vary considerably just by chance alone. It also follows that, although there is not any inherent bias in the procedure of random sampling, it is perfectly possible that any one sample may differ considerably from the population from which it was drawn. This means that the sample can sometimes be unrepresentative of the wider population and provide a misleading snapshot of the bigger picture. This phenomenon is known as sampling error and it is looked at again later in the chapter. It is also a theme that is revisited in Chapter 14, 'Deciding on differences'.

Systematic sampling

Although random sampling is an attractive means of choosing a sample, it is not always the most convenient method. For example, if you wished to find out about consumers' brand preferences for various goods, it would make more sense to stand outside a few selected supermarkets and shops and ask people as they come out. Three further sampling techniques that are commonly used are systematic, stratified and cluster sampling. This section looks at systematic sampling.

Most street or home-based polls are based on some sort of systematic sampling method. If the polling agency is aiming to carry out, say, a 10% poll of a street, the interviewer may be instructed to knock on every tenth door or to stop every tenth passer-by.

An obvious advantage of systematic sampling is that it should guarantee an even spread of representation across the sampling frame. However, there are certain dangers inherent in systematic sampling, as the next exercise should reveal.

Exercise 7.3 Systematic bias

What biases might the following sampling procedures suffer from?

▶ Estimating yearly sales of a shop over a ten-year period by sampling sales figures systematically every six months.

▶ Estimating a household's monthly expenditure by sampling their bills and payments on the first week of every month.

▶ Estimating the absentee rates of a firm by sampling their absentee rates every 5 working days.

[Comments below]

All these examples carry the dangers of picking up a cyclical pattern in the population data and repeating it again and again.

For example, sampling sales every six months will pick up seasonal fluctuations. So, if the first month were, say, June then every other month sampled would be a December, in which case the untypical pre-Christmas rush would be picked up each year. Taking the second example, household expenditure patterns are never uniform over a month. Many people arrange that standing orders are paid out at the beginning of the month and they may hang on until the benefit payment or wage packet arrives before taking on a major shopping expedition. Finally, the third example – if a survey on absenteeism is carried out on the same day each week, certain biases such as 'Monday morning syndrome' or 'Friday afternoon syndrome' may distort the findings.

These examples of systematic bias are fairly obvious and it wouldn't be difficult to set the repeating samples cycle to avoid them. However, the danger with systematic sampling is that you may pick up an underlying cycle in the population that you weren't aware of, and so introduce bias unknowingly. There is no easy remedy here except to use a bit of common sense and imagination.

Stratified and cluster sampling

To end this section, it is worth mentioning two further sampling techniques – stratified and cluster sampling.

Stratified sampling: This involves partitioning the sampling frame into two or more separate 'strata' or layers and then sampling separately from each stratum, thereby ensuring that each group is fully represented in the survey. The numbers of people or items chosen to represent each stratum should match the proportions that occur in the overall population. This is a widely used technique and it requires that the strata are easily identifiable in the population. For example, the designers of a survey on smoking may wish to identify three key categories – non-smokers, light smokers and heavy smokers – and try to ensure that each of these strata is represented within their sample in the same proportions that occur in the overall population. The term 'random stratified sampling' is where the sampling frame is firstly stratified and then the items are sampled randomly from each stratum.

Cluster sampling: Suppose that a sample is required for interview from a large population such as all the inhabitants of the UK. Clearly the people selected for the sample may be very widely spread geographically and the costs of travelling to each separate address would be huge. Cluster sampling is a way of reducing the time and cost involved by restricting the sample to a smaller number of geographical areas. The population in each of these areas is a cluster and the sub-samples from each cluster are combined to produce the final sample.

Nugget: when the polls went wrong

In the US presidential election of 1948 the two frontrunners were Harry Truman (Democrat) and Thomas Dewey (Republican). On the eve of the election there was widespread agreement among the main polling agencies that Dewey was going to win. Indeed, so certain were they about the outcome that several newspapers carried banner headlines proclaiming Truman's defeat before the returns were in. For example, the *Chicago Tribune*, a pro-Republican newspaper, printed 'DEWEY DEFEATS TRUMAN' on election night as its headline for the following day. In the event they were wrong and Truman won by a slim majority (final result, Truman 49.6%, Dewey 45.1%).

The failure brought private and academic researchers together to consider why they had been so overconfident in their predictions of the outcome. Here are some of the lessons they learned:

▶ There was a very late swing in the final hours before voting which the polling agencies failed to spot.

▶ In some cases, telephone polling was used, which introduced a bias towards better-off (and therefore Republican) voters.

▶ They assumed that the votes of the 'undecideds' would split along similar lines to those of the 'decideds'.

▶ They had no good mechanism for working out who would vote and who would stay at home on voting day.

▶ Today, apart from the occasional blip, polling agencies are able to predict population behaviour with remarkable accuracy.

Human error and sampling error

I reckon the light in my 'fridge' must stay on all the time. I've checked hundreds of times and every time I open the door to have a look it is still on!

One of the dilemmas for any researcher is that he or she never really knows the extent to which their research observations affect or disturb the objects or people being observed. As with the 'opening the "fridge" door' example above, all surveys involve opening some sort of door on people's lives. Observers participate in this process and can never fully know the extent to which their presence has affected the phenomena that are being presented to them. All they can do is attempt to monitor their findings against those collected in a different way and then try to account for any differences. Most of all, the observer needs to be sensitive to these questions and make every attempt to tread lightly.

There are two main types of error that surround sampling. The first is the sort introduced by a badly designed sampling procedure, and some examples of these have been given earlier in the chapter. This might be called 'human error'. Sometimes error is deliberately and unfairly introduced into the sampling process in order to distort the findings. Examples of this are the original polls which were used to compile the pop music charts. The first singles chart, known as the Record Hit Parade, was released in 1952 and was published by the *New Musical Express*. (Before that, the charts were compiled from sheet music sales.)

In those early days, a small number of record shops were selected and sampled for their record sales over the previous week. The problem was that the sample size was small and, once the identity of a 'chart shop' became known, agents would descend with wads of fivers to buy particular records and thereby bump up their chart ratings. This problem was tackled by increasing the number of 'chart shops', and the Gallup polling agency responsible for compiling the charts subsequently sampled from around 1000, or roughly one in four of all music outlets.

Phone-in polls were a popular means by which some radio stations compiled record charts, but these are even more open to distortion than sampling sales from record outlets. In 1986, when Capital Radio decided to compile the all-time Top 100 records, based on a listeners' phone-in poll, five Bros records featured in the chart. This probably says something about the promotional zeal of the listening 'Brosettes' and the fact that a phone call or ten costs less than the price of a single, particularly if mummy and daddy are footing the telephone bill.

Let us now turn to the second form of error, known as 'sampling error'. Due to natural variation, all sampling is inherently prone to variation that is an inevitable part of the random sampling process. It is this type of error that was first mentioned earlier in this chapter in the section headed 'Sampling variation'. Although sampling error cannot be eliminated, it is possible to quantify how much variation to expect under different conditions, and this allows us to gain some measure of the confidence with which we can make predictions about the population.

For example, a manufacturer sells a popular brand of baked beans marked with an average contents weight of 415g. As part of its quality control procedures, the company regularly weighs a sample of 100 tins to ensure that they are of the required weight. Due to natural variation, the weight of the contents of each tin will inevitably vary. So, to ensure that they don't contravene the weights & measures legislation, the manufacturer will aim for an average weight that is slightly greater than 415g – perhaps something like 425g.

Let's say that in a particular sample, the lightest weighing was 414g and the heaviest was 434g. The sample mean will lie somewhere between these two extremes – perhaps something like 423g. Because it is a sample mean of quite a large sample (in this case, 100 tins), it provides a much closer estimate of the average weight of the contents of a tin than you'd get by weighing a single tin. On the basis of knowing the mean and spread of this sample, it's possible to come up with what is termed a 'confidence interval' for the average weight of the contents of a tin of beans. Exactly how a confidence interval is calculated is beyond the scope of this book but it might be stated in the following form (the numbers used here are purely illustrative).

We are 95% confident that the average weight of the contents of a tin is between 419 and 427g.

What it means in practice is that if the manufacturer were to sample the weights of 100 tins of beans every day for a month, they would expect that 95% of the sample means would lie within this range. This 419–427g interval can be described as a 95% confidence interval and this sort of statement (where the averages are reduced to an interval rather a single value) is known as an 'interval estimate'. Such a statement is particularly useful as a way of thinking through the worst-case and the best-case scenarios for a sample average.

TWO EXPERIMENTS

In the previous section, you saw that it was possible to come up with a confidence interval for the average weight of the contents of a tin of beans, based on taking a sample of size 100. In this final section, you are asked to think about what would be the effect on the confidence interval for a smaller or larger sample. For example, what sort of confidence interval might you expect to get with a tiny sample size of, say, 5 tins or with a very large sample size of, say, 1000 tins? With this in mind, have a go at Exercise 7.4 now.

Exercise 7.4 Spot the connection

The width of the confidence interval is related to the sample size. How would you expect these two factors to be related?

[Comments below]

The larger the sample you take, the greater your confidence will be that the sample result is accurate, so therefore the narrower will be your confidence interval. This may appear to be slightly counterintuitive (the notion of high confidence being linked to 'narrow' confidence) but it may be helpful to think of the confidence interval as a 'tolerance' – a low error tolerance is more comfortably associated with a high degree of confidence in the population estimate. The best way to be convinced of this relationship between confidence and sample size is to do a simple sampling experiment based on samples of different size. If you then plot the sample means you will find that the means which were calculated from the smaller samples are spread out more widely than the means taken from large samples. This basic idea is first illustrated with a simple example using dice. The same point is then made using a larger sample, this time with random numbers.

This first experiment is about finding the mean score when a die is tossed more than once. Regardless of how many times the die is tossed, the mean score that you would expect to get is around 3.5. (This is because the six outcomes 1, 2, 3, 4, 5 and 6 are equally likely to occur and the mean of these six numbers is 3.5.) Of course the means will not always equal 3.5 – sometimes they will be more and sometimes less. Now imagine two dice, one black and the other white. The black die is rolled twice and the mean of the two scores taken. The white die is rolled six times and the mean of the six scores is taken. In which case might you expect the mean to be closer to 3.5? Will it always be so?

 Exercise 7.5 Dice games

Find a die and carry out a similar experiment to the one described below – this will only take you a few minutes and should ensure that you grasp the essential point of it. Check that the same outcome emerges as the one described here.

a Roll the black die twice and calculate the mean: e.g. 5, 1 → mean = 3.

b Roll the white die six times and calculate the mean: e.g. 6, 2, 4, 3, 3, 1 → mean = 3.17.

c Note which of the two answers is closer to 3.5.

d Repeat the experiment several times. Overall, which die, the black or the white, tends to 'win' – namely to lie closer to 3.5?

[Comments below]

You probably found that the white die, based on a mean from a larger sample size, tended to win this particular dice game. This result appears to confirm the earlier claim, that 'the larger the sample you take the narrower will be your confidence interval'. However, this was a very crude and small-scale experiment. In Table 7.1, the same basic exercise has been done, but on a rather larger scale using random number tables.

Table 7.1 Investigating sample means using different sample sizes

Random X	Means (n = 2)	Means (n = 3)	Means (n = 5)	Means (n = 10)
4				
9	6.5			
4	6.5	5.7		
8	6.0	7.0		
7	7.5	6.3	6.4	
5	6.0	6.7	6.6	
2	3.5	4.7	5.2	
8	5.0	5.0	6.0	
0	4.0	3.3	4.4	
2	1.0	3.3	3.4	4.9
6	4.0	2.7	3.6	5.1
2	4.0	3.3	3.6	4.4
0	1.0	2.7	2.0	4.0
5	2.5	2.3	3.0	3.7
8	6.5	4.3	4.2	3.8
8	8.0	7.0	4.6	4.1
7	7.5	7.7	5.6	4.6
8	7.5	7.7	7.2	4.6
2	5.0	5.7	6.6	4.8
2	2.0	4.0	5.4	4.8
1	1.5	1.7	4.0	4.3
4	2.5	2.3	3.4	4.5
7	5.5	4.0	3.2	5.2
0	3.5	3.7	2.8	4.7
4	2.0	3.7	3.2	4.3
1	2.5	1.7	3.2	3.6
8	4.5	4.3	4.0	3.7
5	6.5	4.7	3.6	3.4
1	3.0	4.7	3.8	3.3
9	5.0	5.0	4.8	4.0
1	5.0	3.7	4.8	4.0
7	4.0	5.7	4.6	4.3
8	7.5	5.3	5.2	4.4

	Random X	Means (n = 2)	Means (n = 3)	Means (n = 5)	Means (n = 10)
	8	8.0	7.7	6.6	5.2
	9	8.5	8.3	6.6	5.7
	4	6.5	7.0	7.2	6.0
	5	4.5	6.0	6.8	5.7
	8	6.5	5.7	6.8	6.0
	6	7.0	6.3	6.4	6.5
	9	7.5	7.7	6.4	6.5
Max	9	8.5	8.3	7.2	6.5
Min	0	1.0	1.7	2.0	3.3
Range	9	7.5	6.6	5.2	3.2

The calculations in Table 7.1 were performed on a spreadsheet, which took a lot of the hard graft out of the activity! The column headed 'Random X' contains the first 40 random digits in row 07 of Figure 7.2. The first figure in column 2 shows the mean of the first two of these random numbers (i.e. 6.5 is the mean of 4 and 9). The second figure in column 2 shows the mean of the second and third random numbers, and so on. The figures in the other columns show the result of averaging the random numbers three at a time, five at a time and ten at a time.

At the bottom of the table are given the minimum value, maximum value and range of each set of sample means. (Note that Range = Max – Min.)

Notice how the mean values in column 2 vary enormously from a minimum of 1.0 to a maximum of 8.5, giving a range of 7.5. However, as the sample size is increased to 3, then 5 and finally 10, this spread of the mean values is progressively reduced (to a range of 6.6, 5.2 and 3.2, respectively). Once again, this confirms the 'central finding', namely that means calculated from samples with larger sample sizes tend to lie closer to the true mean than means calculated from samples with smaller sample sizes. This is an important result in statistics and is the central plank in what is known as the 'central limit theorem'. In statistical terms, it is the key element in the Keith Parchment story that started this chapter off, namely that confidence about the results of sampling is increased as sample size is increased.

Key ideas

► The distinctions between the terms population, sampling frame and sample.

► The idea of random sampling.

  Have a look at the random number tables to give these ideas a practical context.

► Sampling variation and alternative sampling techniques – systematic, stratified and cluster sampling.

► The distinctions between human error and sampling error.

► The ideas of a confidence interval and an interval estimate.

► The 'central limit theorem', namely that means calculated from samples with larger sample sizes tend to lie closer to the true mean than means calculated from samples with smaller sample sizes.

To find out more via a series of videos, please download our free app, Teach Yourself Library, *from the App Store or Google Play.*

8

Collecting information

In this chapter you will learn:

▶ *where to find data that you might use in an investigation*

▶ *how to design an experiment to collect your own data*

▶ *some dos and don'ts about questionnaire design.*

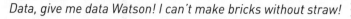

Data, give me data Watson! I can't make bricks without straw!

Sherlock Holmes

In Chapter 1, four key stages of a statistical investigation, known as the 'PCAI' cycle, were listed as follows:

- ▶ P   Stage 1   pose a question
- ▶ C   Stage 2   collect relevant data
- ▶ A   Stage 3   analyse the data
- ▶ I   Stage 4   interpret the results

We now turn to the 'C' stage of the PCAI cycle: collecting relevant data. In any statistical work, the relevance of data is determined by whether or not it will help you to answer your central question. It cannot be stressed enough that the more purposeful and clearly formulated your central question is (stage 'P'), the easier it will be to make sensible decisions at the 'C', 'A' and 'I' stages later on.

The first guiding principle at the 'C' stage is to remember that data generation costs time and money, so avoid re-inventing the wheel. In other words, before you rush out and try to generate your own data, make sure that someone else hasn't already done it for you. Data which have already been collected by someone else and are awaiting your attention are known as *secondary source* data or sometimes simply as *secondary* data. In the absence of suitable secondary source data, you may have to carry out an experiment or survey yourself, in which case this would involve you in generating *primary source* data (also known as *primary* data). Libraries and the Internet are full of valuable secondary source data and these are probably the first places you should look.

Secondary source data

Historically, the term 'statistics' derives from 'state arithmetic'. For thousands of years, rulers and governments have felt the need to measure and monitor the doings of their citizens. From raising revenue in the form of taxes to counting heads and weapons in times of war, the state has always been the major driving force in collecting and publishing statistical information of all kinds. And, as you will see from this section, they are still at it! Just a few of the most useful secondary sources of data are described here, most of which are from government sources.

Let us assume that you have identified a particular question for investigation. As has already been suggested, the first task is to check whether any suitable secondary source data are available.

A good starting point is to track down the government body concerned with collecting and presenting national statistics. In the UK this role is carried out by the Office for National Statistics (ONS) and its website is: www.ons.gov.uk In the US the corresponding body is the statistics branch of USA.gov: www.usa.gov/statistics

The ONS is responsible for collecting and publishing statistics related to the economy, population and society at national, regional and local levels. It also conducts the census in England and Wales every ten years. Here are just some of the areas on which it provides helpful and interesting data.

Table 8.1 An indication of the many datasets provided by the ONS

General area	Examples
Business	The construction industry
	The IT industry
	The retail industry
	Tourism
The Economy	Economic output and productivity
	Environmental accounts
	Inflation and price indices
	National and regional accounts
Employment and the Labour Market	People in work
	People not in work
People, Population and Community	Births, deaths and marriages
	Crime and justice
	Health and social care
	Population and migration

A vital source of important data collected and made available by the ONS is the Census of Population. The census takes place every ten years (for example, in 2001, 2011, 2021, etc.) and is based on a restricted amount of information about personal details and the nature of households of all citizens of the UK, as well as details of international migration. It is used as a source of reliable statistics for small groups and areas throughout the country so that government and others can plan housing, health, transport, education and other services. The 2011 census data were collected from some 26.4 million households and made generally available in 2015. Local area statistics from the census were published separately for each county and all the key results of each county are also available in machine-readable format.

One frustration with secondary data is that they are rarely in exactly the form that you want. You may think that you have collected information on gross earnings in the UK only to read the small print and discover that the figures refer to Great Britain only or that part-time workers have been excluded and you need the inclusive figure for purposes of comparison. Fortunately there is a simple remedy for such problems – it is called 'hard graft'! You need to be prepared to put in the time and effort checking out all the small print of any secondary data that you propose to use and making sure that they really are the figures that you want. As often as not you will need to chase up some additional data from elsewhere.

Data accuracy

All data that have been collected by measurement of some sort, whether secondary source or primary source, are inaccurate. This is so for a variety of reasons. First, no measuring tool is perfectly accurate. If the information has derived from physical measurement, the ruler or weighing scales or thermometer, etc. can only be read off to a certain degree of accuracy. The following exercise makes this point.

Exercise 8.1 Paper thin

A block of 60 sheets of writing paper is measured at ½ cm in thickness. How thick is each sheet?

[Comments below]

Now ½ cm = 5 mm. So, the appropriate calculation seems to be:

Thickness of 60 sheets = 5 mm
Thickness of one sheet = 5 ÷ 60 = 0.083 333 3 mm

However, how confident are you of this result to eight figures of accuracy? Probably not at all! There is no clear indication in the question as to the measuring method used and the degree of accuracy of the original measurement of ½ cm. A difficulty with expressing numbers as fractions like this is that it is impossible to apply the normal conventions of accuracy. If the original measurement had been expressed as a decimal number (either 0.5 cm or 5 mm) it would be reasonable to assume an accuracy of 1 mm. In other words, the block of paper was between 4.5 and 5.5 mm in thickness. (Note that if the measurement had been carried out to an accuracy of, say, 0.1 mm, the result would have been written as either 0.50 cm or 5.0 mm.) If we take these two maximum and minimum values separately, we get the following range of values for the thickness of a single sheet.

Minimum thickness of one sheet = 4.5 ÷ 60 = 0.075 mm
Maximum thickness of one sheet = 5.5 ÷ 60 = 0.091 666 7 mm

An awareness of the range of possible values – from smallest to greatest – emphasizes the nonsense of quoting the original answer to eight figures. In this and similar examples, it is sensible to *round* the answer to an appropriate degree of accuracy. Here an answer of 'about 0.08 mm', or even 'about 0.1 mm', would seem to be an appropriate estimate for the thickness of a single sheet.

Nugget: spurious precision

Whenever you are given an implausibly precise statistic, you would be correct to smell a rat and want to investigate whether it might be bogus. The made-up example below is so bogus you can actually smell a dinosaur!

When a visitor at the Natural History Museum in London asked an attendant how old a particular dinosaur was, she was told 200 million years, 8 months, 1 week and 2 days. 'But how can you be so certain?' she asked.

A second source of data inaccuracy is human error. This may be due to the researcher misreading the instrument or perhaps miscoding or miscounting the category responses to a survey. An additional dimension to the incidence of 'human error' in survey work is that respondents may, either knowingly or unknowingly, be providing false information. There have been several well-documented historical examples of this in the field of anthropology, where European researchers have visited isolated, 'primitive' island communities and reported complex and bizarre sexual rituals, only for subsequent researchers to discover that the 'primitive islanders' were, in fact, engaging in a jolly leg-pull exercise and had been making up whoppers for their own amusement!

Sometimes it is possible to identify and even quantify respondent error in follow-up surveys. For example, in the United States, the Bureau of the Census has a policy of resurveying a sample of the original respondents for verification. One of the questions asked of unemployed respondents is to state the number of weeks that they had been unemployed. In the follow-up interview, say 5 weeks later, one might expect those respondents who are still unemployed to produce an answer to this same question that is 5 weeks longer than their earlier answer. Some hope! In the event, only about one-quarter give consistent responses. Of the other three-quarters, some 'gain' weeks of unemployment between interviews and others 'lose' weeks of unemployment. The problem for the researcher is deciding which (if any) of these responses to believe. It is also worth noting that this example reveals a high likelihood of widespread inaccuracy in the responses from most respondents to most questions. Let's not fall into the trap of assuming that just because someone tells you something in an 'interview' situation that they really know the answer, or that their reply is necessarily true!

Primary source data

It may be that your research question is concerned with an issue that is too specific or too local for there to be suitable secondary source data available, in which case you may decide to collect your own data. This may require accurate measurement that involves carrying out some sort of scientific experiment, or perhaps designing a questionnaire and conducting a sample survey. In this section we will look at two of the key issues in this area: experimental design and questionnaire design.

EXPERIMENTAL DESIGN

Some years ago, the parents of two teenagers, both of whom had committed suicide, sued the rock group Judas Priest and their record company, CBS records. The parents alleged that the deaths were partly due to subliminal messages (the words 'Do it') concealed in one of the group's albums, *Stained Class*. In the event, the group and their record company were cleared of the charges by a Los Angeles court, but the case raised considerable interest in the whole area of the effects of subliminal messages, both in relation to advertising and for educational purposes. A number of commercial self-help audio recordings have been produced containing subliminal messages against a background of relaxing music, all designed to help people lose weight, give up smoking or even improve their memory. Over a five-week period, a group of people listened every day to a recording that contained subliminal suggestions for improving memory. Half felt that their memory had improved. Quite impressive, you may feel! However, the researchers also asked a second group to listen to a recording that they *thought* contained messages but in fact did not. Well, guess what? Roughly half of the second group reported the same thing. So what is going on here? According to the report of the British Psychological Society (*Subliminal messages in recorded auditory tapes and other unconscious learning phenomena*), the explanation may be as follows:

When someone expects something to be helpful, there can be a positive outcome, even if the product itself is useless. People tend to avoid the embarrassment of confessing to themselves that a financial investment has been wasted.

Exercise 8.2 Cure or con?

Suppose that a drug company has claimed to have invented a cure for the common cold. They have tried it out on ten cold-suffering volunteers and, indeed, all ten got better. Why might you not be convinced by their conclusions that the drug really works? How could you improve on the experimental design in order to give the drug a fairer test?

[Comments below]

Rest assured that drug companies would never get away with this experiment as there are strict rules to be observed in the manufacture and testing of drugs before they are licensed and made available to the general public. Here are some of the experimental weaknesses in the example above.

▶ **Who says they are better?** – The states of being 'well' or 'ill' or 'better' or 'worse' are often highly subjective. Who is making the health judgement on these subjects and do they have a vested interest in the outcome? It is possible that the drug has simply suppressed the symptoms (headache, sneezing, watering eyes, etc.) but the virus is not eliminated.

▶ **Sample size** – A sample of only ten subjects is much too small.

▶ **Timescale** – We don't know if recovery took place over a period of hours, days or years.

▶ **Controls** – There is nothing to compare with here. A key question is whether the subjects would have recovered anyway without this treatment.

▶ **The placebo effect** – This is sometimes also known as the 'feel good' factor. Particularly in the area of health, people often respond positively to psychological influences – simply as a result of swallowing a pill or drinking some nasty-tasting medicine. So the psychological benefits of being given treatment may be obscuring the chemical merits or demerits of the drug.

Proper drug testing is based on an experimental design which attempts to meet these objections.

On the question of 'Who says they are better?', there need to be, where possible, objective measurable criteria (blood or urine tests, body temperature checks and so on) to assess whether or not the subject's health has indeed improved after taking the drug.

A sample size of several hundred or several thousand would be required in order to test a drug rigorously. This is necessary for two reasons. First, a reasonably large sample size is needed in order to carry out subsequent analysis of tests of significance at the level of significance required. Second, when licensed, the drug will not be administered to a homogeneous collection of patients, but rather to people of a wide variety of ages, body weights, medical backgrounds and so on. It may also be offered to pregnant women. Thus, testing needs to be on a scale that is sufficient to establish with some degree of certainty whether or not the drug is safe for this range of potential users. Testing may also be aimed at establishing effective and safe dosage levels for adults and children separately.

Nugget: deciding on sample size

A question that any researcher will want to ask is, 'How large a sample size should I choose?' There are three key factors to consider here.

a *The level of precision required.* If you are investigating, say, the effects of brain surgery, your conclusions will need to be more precise than a survey on something mundane such as washing powder. The greater the precision required the larger a sample size you will need.

b *The confidence level you choose.* If, in your research, you decide on a confidence level of, say, 95%, this means that you can expect 95% of any similar samples that you might select to share your finding. Choosing a confidence level of 99% would mean that you can expect 99% of any further samples that you might select to share your finding. The greater the level of confidence you wish to aim for the larger the sample size you need to use.

c *Degree of variability of the data.* Some data items possess a lot of inherent variability – for example, if you select and accurately weigh a sample of similarly manufactured bags of potato crisps you'll find that their weights will vary quite a lot. Other precision-manufactured items, such as grommets or screws, will vary very little in size from one to the other. Items such as bags of crisps that possess high natural variability will require larger samples sizes if you are to be able to draw satisfactory conclusions from an investigation into their sizes.

The timescale over which the drug appeared to be effective also needs to be determined. This is important both to check that the drug really is working (people tend to recover on their own eventually if left untreated) and also to fine-tune the period over which the drug needs to be taken (there is no need to keep taking a drug long after it has cured you).

Normal practice with drug testing is to create two similar (randomized) groups of subjects. The *experimental group* will be given the drug being tested, while the *control group* is given a placebo. The placebo will take the form of a completely harmless and neutral 'look-alike' form of treatment (perhaps an injection of distilled water or an identical pill that contains no active drug). It is essential that the subjects are not told which group they are in, otherwise the 'placebo' effect will operate in the experimental group and not in the control group. This sort of experiment, where the subjects are not told whether they are in the experimental or the control group, is known as a *blind experiment* for obvious reasons. Of course, in certain situations it is possible that the researcher's knowledge of which group each subject has been allocated to affects his or her behaviour in a way that influences the health of the subjects. In order to avoid this possibility, it may be a good idea to set up a *double-blind experiment* where neither the experimenter nor the subjects know who is in which group.

Blind trials may be methodologically sound, but they do raise certain moral dilemmas. Consider the situation where a new drug has been discovered which appears to cure a fatal illness – perhaps terminal cancer or AIDS. The volunteers who offer themselves to be tested are all desperate. This drug could be their last chance and it may seem a cruel exploitation of their plight merely to inject half of them with distilled water in the cause of adopting correct scientific procedure.

QUESTIONNAIRE DESIGN

Exercise 8.3 A questionable questionnaire

The questionnaire below, surveying people's health practices and attitudes, is based on examples given in the book *Surveys in Social Research*, by D.A. de Vaus (2002). Each question is something of a lemon in the way it is worded. Have a go at answering the questionnaire and then give some thought to the sorts of biases and inaccuracies it might produce.

A survey on health

1 How healthy are you?

2 Are the health practices in your household run on matriarchal or patriarchal lines?

3 Has it happened to you that over a long period of time, when you neither practised abstinence nor used birth control, you did not conceive?

4 How often do your parents visit the doctor?

5 Do you oppose or favour cutting health spending, even if cuts threaten the health of children and pensioners?

6 Do you agree or disagree with the following statement? 'Abortions after 28 weeks should not be decriminalized.'

7 Do you agree or disagree with the government's policy on the funding of medical training?

8 Have you ever murdered your grandmother?

[Comments below]

As you will have discovered, this is a pretty hopeless questionnaire but there are some important general principles that can be learned from it.

1 There is no way of knowing how to answer a question like this. There needs to be included either a clear set of words to choose from or a scale of numbers on which to rate your perception of how healthy you feel.

2 Don't ask questions that people won't understand.

3 Simplify the wording and sentence structure so that the question is clear and unambiguous without being trite or patronizing. Don't use undefined time spans (such as 'a long period of time') and avoid using double negatives.

4 Avoid double-barrelled questions like this where two (or more) individuals have been collapsed into one category (in this case, 'parents'). Also, where a quantitative answer is required, give the respondent a set of categories to choose from (e.g. 'every day', 'roughly once a week', etc.). Finally, avoid gathering information from someone on behalf of someone else. They probably won't know the exact details and anyway it is none of their business.

5 This is a leading question that is pushing for a particular answer and will produce a strong degree of bias in the responses.

6 Again, avoid double negatives as they are hard to understand. This could be reworded as 'Abortions after 28 weeks should be legalized.'

7 Most people will not know what the government's policy is on the funding of medical training, so don't ask questions that people don't know about without providing further information.

8 This could be classed as a direct question on a rather sensitive issue. If you really must ask this question, there are ways of asking it more tactfully. According to de Vaus, there are four basic gambits.

   a The 'casual' approach – 'Do you happen to have murdered your grandmother?'

b The 'numbered card' approach – 'Will you please read off the number on this card which corresponds with what became of your grandmother?'

c The 'everybody' approach – 'As you know, many people have been killing their grandmothers these days. Do you happen to have killed yours?'

d The 'other people' approach – 'Do you know any people who have murdered their grandmothers?' Pause for a reply and then ask, 'Do you happen to be one of them?'

There are several other aspects to designing a good questionnaire that have not emerged from this one. First and foremost, only ask questions that you want and need to know the answers to and that will provide you with information which you intend to use. Most people have only a limited capacity for filling in forms and answering questions, so don't use up their time and goodwill asking unnecessary questions. Second, you will need to decide whether to collect quantitative or qualitative data or a combination of the two. In making this choice, it is helpful to bear in mind how you intend, subsequently, to process the information that you collect. Qualitative information will be difficult to codify and summarize but it may allow options and opinions to be expressed that you would never otherwise hear. Quantitative data are easier to process but have the disadvantage that the range of responses is restricted to the ones you thought of when designing the questionnaire. A helpful solution to this dilemma is to offer a set of pre-defined categories but allow the respondent to add a final option of their own in the space provided, thus:

'Other, please specify' _____

Also, with questions where opinions are being sought, there may be issues on which people will have no opinion, so the option of recording 'don't know' or 'no opinion' should be offered in the range of choices.

Let us now return to an earlier suggestion that double negatives could and should be avoided by adopting a simpler wording. This is not as simple as it sounds, as the next exercise will reveal.

Tick the boxes below to indicate whether or not you agree or disagree with these statements.

	Yes	No
a I do not dislike my doctor	☐	☐
b I like my doctor	☐	☐

[Comments below]

Two points emerge from this exercise. First, rewording a double negative may fundamentally alter the meaning of the original question. In this particular example, Statement b is a much stronger vote in favour of your doctor than Statement a. Second, you may have observed that the options 'Yes' and 'No' are not entirely neutral words.

In conclusion, be prepared to pilot your questionnaire (try it out on a few friends first) and redraft it several times before embarking on the major survey. The layout needs to be thought about and improvements can usually be made in its general presentation. Think about the order of the questions that you have asked – is it logical and sensible? And finally, maintain respect and confidentiality for the responders and any data that you subsequently collect from them. You have both a moral and a legal duty to do so!

To end the chapter, the final exercise gives you an opportunity to put some of these ideas of good questionnaire design into practice.

Exercise 8.5 It's your verdict!

Under the banner headline 'We've had ENOUGH!', a tabloid newspaper produced the following questionnaire. Make a note of any criticisms of the way it was designed.

IT'S YOUR VERDICT

1 I believe that capital punishment should be brought back for the following categories of murder:
Children ☐   Police ☐   Terrorism ☐   All murders ☐

2 Life sentences for serious crimes such as murder and rape should carry a minimum term of:
20 years ☐   25 years ☐

3 The prosecution should have the right of appeal against sentences they consider to be too lenient. ☐

*Tick the boxes of those statements you agree with and then post the coupon to:

VIOLENT BRITAIN, *Daily Star*, 33 St. Bride St., London EC4A 4AY

[Comments below]

Comments on exercises

▶ **Exercise 8.5**

The main criticism to be made of this questionnaire is the remarkable degree of restriction of the response categories offered. For example, question 2 offers only two heavy terms of sentence and there simply isn't a box for someone who would like to be 'soft on murderers and rapists' and believes that longer prison sentences won't necessarily solve the problem of violent crime. Also, the responses available in question 1 provide no opportunity for someone to express a view if they don't believe in the death penalty. Second, the context of the survey is highly emotive and biased, being placed within headlines such as 'We've had enough', 'My mother's killer runs free' and 'Hang the gunmen'. Perhaps it is not surprising that, from the 40 000 readers who took the trouble to complete and post the coupon, the following results were recorded:

> favouring restoring capital punishment for murder: 86.33%
> favouring a 25-year minimum sentence for serious crimes of violence: 92%
> favouring the right of appeal by the prosecution against sentences they consider to be too lenient: 95.58%

It is interesting to contrast these findings with those of more reputable polling agencies (such as Marplan and the Prison Reform Trust, for example) who consistently report that most people, including victims of violent crime, would prefer to see compensation to victims and a system of community service rather than longer prison sentences.

Key ideas

▶ *Secondary* sources of data have already been collected by someone else, whereas *primary* sources of data you collect yourself.

▶ An important and valuable online source of secondary data in the UK can be found at the website of The Office for National Statistics (ONS): www.ons.gov.uk

▶ There are two main types of error: those resulting from the inevitable limitations of the measuring tool used and human error.

▶ There are two main issues in the area of collecting primary source data: experimental design and questionnaire design.

To find out more via a series of videos, please download our free app, Teach Yourself Library, *from the App Store or Google Play.*

9

Spreadsheets to the rescue

In this chapter you will learn:

▶ *why spreadsheets are helpful in everyday situations*

▶ *some useful spreadsheet commands for performing calculations and summarizing data.*

If possible you should try to obtain access to a spreadsheet application on a computer for this chapter.

These days, data handling is increasingly done on a computer and the most common application for this purpose is a spreadsheet. If you have access to a computer with a word processor, the chances are that it already has a spreadsheet application available to you as part of a wider 'office software' package. The most commonly used spreadsheet application is Excel, which is part of Microsoft Office (popular but expensive). Other packages include Star Office and Calc (a free spreadsheet package that is part of 'Open Office', available via OpenOffice. org). Also popular is LibreOffice Calc. You can also freely access the web-based package Google Spreadsheets (just type 'Google Spreadsheets' into Google and follow the links provided).

If you want to make a start with *using* spreadsheets, it isn't enough simply to read the words below – you really need to be actively involved. That means firing up a spreadsheet application, actually keying in data and trying out the various commands that are described.

Please note that, in order to encourage you to do this, I have endeavoured to reduce the burden of inputting the data by providing fairly small data sets. However, please don't infer from this that spreadsheets are useful only for small data sets. Nothing could be further from the truth! One of the great strengths of spreadsheets is that, whether your columns of data contained 10 or 100 or 1000 items of data, the commands would have been equally simple to enter and the time taken for your spreadsheet to make the calculation would still be only fractions of a second.

Nugget: saving and backing up data

Saving and backing up your data should be at the top of your computer list of priorities. Failing to do so risks losing your data. And it will happen, so please don't think that you don't have to worry about it!

What is a spreadsheet?

A spreadsheet is a computer tool that provides a way of laying out data in rows and columns on the screen. The rows are numbered 1, 2, 3, etc. while the columns are labelled with letters, A, B, C, etc. Typically, a spreadsheet might look something like the grid in Figure 9.1, except that rather more rows and columns are visible on the screen at any one time.

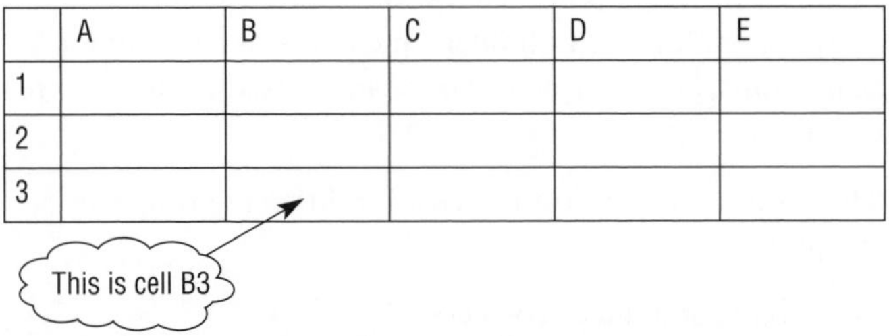

	A	B	C	D	E
1					
2					
3					

This is cell B3

Figure 9.1 Part of a spreadsheet

Each 'cell' is a location where information can be stored. Cells are identified by their column and row position. For example, cell B3 is indicated in Figure 9.1 and is simply the cell in the third row of column B.

The sort of information that you might wish to put into each cell will normally fall into one of three basic types: numbers, text or formulas.

▶ Numbers – These can be either whole numbers or decimals (e.g. 7, 120, 6.32, etc.).

▶ Text – These are symbols or words (e.g. the headings of a row or a column of numbers will be entered as text).

▶ Formulas – The real power of a spreadsheet lies in its ability to handle formulas. A cell which contains a formula will display the result of a calculation based on the current values contained in other cells in the spreadsheet (e.g. an average or a total or perhaps something more complicated such as a bill made up from different components). If these other component values are subsequently altered, the formula can be set up so that it automatically recalculates on the basis of the updated values and immediately gives the new result.

Why bother using a spreadsheet?

A spreadsheet is useful for storing and processing data when repeated calculations of a similar nature are required. Next to wordprocessing, a spreadsheet is the most frequently used computer application in business, particularly in the areas of budgeting, stock control and accounting. Spreadsheets are also being used increasingly by householders to help them to solve questions that crop up in their various roles – as shoppers, tax payers, bank account holders, members of community organizations, hobbyists, etc. They can be used to investigate questions such as:

▶ How much will this journey cost for different groups of people?

▶ Is my bank statement correct?

▶ Which of these buys is the best value for money?

▶ What is the calorific content of these various meals?

▶ What would these values look like sorted in order from smallest to biggest?

▶ How can I quickly express all these figures as percentages?

The reason a spreadsheet is such a powerful tool for carrying out repeated calculations is that, once it has been set up properly, you simply perform the first calculation and then a further command will complete all the other calculations automatically. Another advantage of a spreadsheet over pencil and paper is its size. The grid that appears on the screen (part of which was illustrated in Figure 9.1) is actually only a window on a much larger grid. In fact, most spreadsheets have available hundreds of rows and columns, should you need to use them. Movement around the spreadsheet is also quite straightforward: you can use certain keys to move between adjacent cells or to a particular cell location of your choice.

Using a spreadsheet

In this section you will be guided through some simple spreadsheet activities so, if possible, switch on the computer and let's get started.

There are many spreadsheet packages on the market and fortunately their modes of operation have become increasingly similar in recent years. However, your particular package may not work exactly as described here so you may need to be a little creative as you try the activities below.

A very useful feature of any spreadsheet is that it can add columns (or rows) of figures. This is done by entering a formula into an appropriate cell. For most spreadsheets, formulas are created by an entry starting with an equals sign =.

Exercise 9.1 Shopping list

Load the spreadsheet package and, if there isn't already a blank sheet open, create one by clicking on File and selecting New.

Using the mouse, click on cell A1 and type in 'Milk'. Notice that the word appears on the 'formula bar' near the top of the screen. Press Enter (the key may be marked Return or have a bent arrow pointing left) and the word 'Milk' is entered into cell A1.

Using this method, enter the data below into your spreadsheet.

	A	B
1	Milk	0.74
2	Bread	0.88
3	Eggs × 12	2.65

Exercise 9.2 Finding totals using a formula

The formula for adding cell values together is = sum( ). Inside the brackets are entered the cells or cell range whose values are to be added together.

Click in cell B4 and enter: = sum(B1:B3)

Press Enter and the formula in cell B4 produces the sum of the values in cells B1 to B3. Now enter the word 'TOTAL' into A4.

Your spreadsheet should now look like this.

	A	B
1	Milk	0.74
2	Bread	0.88
3	Eggs × 12	2.65
4	TOTAL	4.27

Let's suppose that you gave the shopkeeper £10 to pay for these items. How much change would you expect to get? Again, this is something that the spreadsheet can do easily. Calculations involving adding, subtracting, multiplying and dividing require the use of the appropriate operation keys, respectively marked +, −, * and /. (Note that the letter 'x' cannot be used for multiplication: you must use the asterisk, *). You should find these keys, along with the number keys and =, conveniently located on the numeric keypad on the right-hand side of your keyboard. (These keys are all also available elsewhere on the keyboard, but you may have to hunt them down, which takes time.)

Exercise 9.3 Calculating change from £10

Enter the word 'TENDERED' in cell A5, the number '10' in cell B5 and the word 'CHANGE' in A6. Now enter a formula into B6 which calculates the change.

Remember that the formula must begin with an equals sign and it must calculate the difference between the value in B5 and the value in B4.

[Comments at the end of the chapter]

Let's now try a slightly more complicated shopping list, this time with an extra column showing different quantities. Enter the data below into your spreadsheet, starting with the first entry in cell A8.

	A	B	C	D
	DESCRIPTION	QUANTITY	LIST PRICE (£)	COST (£)
8				
9	Pens black	24	0.87	
10	Folders	15	0.65	
11	Plastic tape	4	1.28	
			TOTAL	

In order to calculate the overall total cost, you must first work out the total cost of each item. For example, the total cost of the black pens is $24 \times £0.87$.

Enter into cell D9 the formula: = B9*C9

This gives a cost, for the pens, of £20.88. You could repeat the same procedure separately for the folders and the plastic tape but there is an easier way, using a powerful spreadsheet feature called 'fill down'. You 'fill down' the formula currently in D9 so that it is copied into D10 and D11. The spreadsheet will automatically update the references for these new cells.

Click on cell D9 (which currently displays 20.88) and release the mouse button. Now move the cursor close to the bottom right-hand corner of cell D9 and you will see the cursor change shape (it may change to a small black cross, for example). With the cursor displaying this new shape, click and drag the mouse to highlight cells D9–D11 and then release the mouse button. The correct costs for the folders (£9.75) and the plastic tape (£5.12) should now be displayed in cells D10 and D11 respectively. Now click on cell D10 and check on the formula bar at the top of the screen that the cell references are correct. Repeat the same procedure for cell D11. You should find that they have indeed automatically updated – magic!

Nugget: cross platform

OK, I'll come clean now – I'm an Apple Mac user and have never really got to grips with a PC. Fortunately, as the years pass, this is less and less of a problem. The degree of compatibility between these two 'platforms' continues to improve over time. Nowhere is this more true than with spreadsheet applications, which allow data to be saved, copied and pasted happily between these two operating systems.

Exercise 9.4 Summing up

With the costs in column D completed, you are now able to calculate the total costs. As before, this requires an appropriate entry in D12 using the = SUM command. Do this now; you should get a total bill of £35.75.

A teacher was working with her class to develop their estimation skills by playing an estimation game. She tested them before and after playing the game and part of the results are given in Table 9.1.

Table 9.1 Part of the results of the estimation investigation

	A	B	C	D
	PUPILS	**Score before**	**Score after**	**Percentage change (%)**
1				
2	Bertie	38	50	
3	Moya	61	66	
4	Erskine	18	59	
5	Liam	34	30	
6	Patsy	44	60	

Exercise 9.5 Calculating percentage

Enter the data from Table 9.1 into a new spreadsheet. By the way, you will find that column A is not wide enough to accommodate all the text. Don't worry about this for the moment. When all the data have been entered, move the cursor to the top of the spreadsheet above cells A1 and B1. As the cursor moves to a position close to the vertical line between the column headings A and B, it changes shape. Click on this vertical line and drag to the right to widen column A. Adjust the width of this column until it accommodates all the text.

Using 'fill down', in Column D calculate the percentage increase or decrease in each pupil's score. Which pupil showed the greatest increase? Did any pupil show a decrease?

[Comments at the end of the chapter]

What else will a spreadsheet do?

This chapter can only really scratch the surface of what can be done with a spreadsheet. As has already been mentioned, the tiny data sets used here are merely illustrative and don't properly reveal the power of a spreadsheet.

Once data have been entered into a spreadsheet, there are many options available for seeking out some of the underlying patterns. For example, columns or rows can be re-ordered or sorted either alphabetically or according to size. As you have already seen, column and row totals can be inserted. Also, you can easily divide one column of figures by numbers in an adjacent column to calculate relative amounts or to convert a set of numbers to percentages.

Spreadsheet commands can include a number of different mathematical functions. First, there are the four operations $+$, $-$, $\times$ and $\div$ which can be found directly on the computer keyboard. In terms of statistical work, there are many other functions contained within the spreadsheet's menu options (or they can be typed in directly, letter by letter). These will enable you to select some or all of the data and find various summaries such as the mean, median, maximum value, standard deviation, and much more, at the touch of a button.

FINDING SUMMARIES

As you saw in Chapter 5, two of the most common averages are the arithmetic mean (often referred to simply as the mean), and the median. Confusingly, the spreadsheet command for the mean uses the word 'average' (which is a general term normally reserved to describe any such summary of a middle value).

In order to explore this, and other spreadsheet commands, let's key in two short data sets. Suppose that you wanted to compare the writing styles of two authors, Jane Austen and J.K. Rowling, by looking at the lengths of their sentences (based on counting the number of words they contain).

To keep things simple, we'll choose just ten sentences randomly from their books, *Pride and Prejudice* and *Harry Potter and the Goblet of Fire* – these are shown in rows 2 to 11 of the spreadsheet screenshot below.

	Austen	Rowling	
1	**Austen**	**Rowling**	
2	50	26	
3	63	12	
4	9	42	
5	10	26	
6	37	23	
7	8	62	
8	33	27	
9	46	80	
10	30	44	
11	34	42	
12	32	38.4	Mean
13	33.5	34.5	Median
14	8	12	Min
15	63	80	Max
16	55	68	Range
17	15	26	Lower Quartile
18	43.75	43.5	Upper Quartile
19	28.75	17.5	IQR
20	18.57118437	20.32076114	St. Deviation
21			

Start by entering the following into a blank spreadsheet.

Enter the column headings 'Austen' and 'Rowling' into cells A1 and B1.

Then enter the remaining values for the sentence lengths for Jane Austen into cells A2:A11 and the corresponding data values for the J.K. Rowling sentences into cells B2:B11.

As you will see shortly, it will be necessary only to calculate the summaries for Jane Austen initially. Once these are completed, it will be a simple matter of filling the formulas across from Column A to Column B to find the corresponding summary values for J.K. Rowling.

Exercise 9.6 Summarizing Austen

Into cell A12 enter = **average(A2:A11)** to see the mean of the Austen data.

Into cell A13 enter = **median(A2:A11)** to see the median of the Austen data.

Into cell A14 enter = **min(A2:A11)** to see the minimum value of the Austen data.

Into cell A15 enter = **max(A2:A11)** to see the maximum value of the Austen data.

Into cell A16 enter = **A15−A14** to see the range of the Austen data.

Into cell A17 enter = **quartile(A2:A11,1)** to see the lower quartile of the Austen data.

Into cell A18 enter = **quartile(A2:A11,3)** to see the upper quartile of the Austen data.

Into cell A19 enter = **A18−A17** to see the inter-quartile range of the Austen data.

Into cell A20 enter = **stdev(A2:A11)** to see the standard deviation of the Austen data.

[Comments below]

The table below summarizes the values produced by these commands.

Values	
32	Mean
33.5	Median
8	Min
63	Max
55	Range
15	Lower quartile
43.75	Upper quartile
28.75	Inter-quartile range
18.57	Standard deviation

Before moving on to work out the Rowling summaries, there are some points worth making about two of these commands.

First, the lower quartile command looks like this.

= quartile(A2:A11,1)

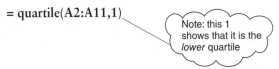

Note: this 1 shows that it is the *lower* quartile

The lower quartile is the value that lies one-quarter of the way through the data, when arranged in order of size; hence, the value 1 in the command. The upper quartile, being three-quarters of the way through the data, shows a '3' in this position.

Second, the spreadsheet actually provides two rather similar commands for standard deviation, = **stdev**() and = **stdevp**(). The one used here, = **stdev**(), is appropriate in situations like this where the standard deviation that you have calculated is being used to make an estimate of the standard deviation of the entire data set (in this case, data taken from all the sentences in *Pride and Prejudice*).

The other command, = **stdevp**(), would be appropriate if you had a set of numbers and you simply wanted to calculate their standard deviation (i.e. you are not concerned with trying to estimate the standard deviation of some wider population).

A full explanation of the distinction between these two formulas is beyond the scope of this book, but essentially the rule is:

▶ Use = **stdev**() when you are dealing with data that represent a sample taken from a wider population and you want to use this calculated value in order to estimate the standard deviation of the wider population.

▶ Use = **stdevp**() when you simply wish to find the standard deviation of a set of numbers (in this case, the 'p' in the command refers to 'population', i.e. you are treating the numbers as the complete population).

Exercise 9.7 Summarizing Rowling

As you saw earlier in the chapter, having entered the commands for calculating the Jane Austen summaries, it is but a flick of the wrist to copy these commands to another column to summarize the Rowling data.

Here are the instructions once more. Select cells A12:A20. Release the mouse button and move the cursor so that it hovers over the bottom right corner of cell A20. When the cursor shape changes (it may become a thin black cross), hold and drag right to cell B20 and then release the mouse button. You should find that all the formulas from Column A have copied across to Column B and automatically updated the relevant cell references so that they now apply to the data for that column.

[Comments follow]

The table below shows the values now displayed in cells A12:B20. (For convenience, some of these numbers have been rounded.)

Values	Display	
32	38.4	Mean
33.5	34.5	Median
8	12	Min
63	80	Max
55	68	Range
15	26	Lower quartile
43.75	43.5	Upper quartile
28.75	17.5	Inter-quartile range
18.57	20.32	Standard deviation

Exercise 9.8 Drawing conclusions

What conclusions might these summaries enable you to draw about the sentence lengths of these two authors?

[Comments below]

The mean and median summaries suggest a fairly similar average length, with Rowling's sentences containing slightly more words, on average.

As far as the spread of the data is concerned, the range and standard deviation summaries both suggest a slightly wider spread with the Rowling than the Austen data. However, overall, a sample size of ten sentences is really too small to make firm conclusions.

Before ending this exploration of spreadsheets, it is worth mentioning that most spreadsheet applications have powerful graphing facilities which allow you to select either all or some of the data and display them as a pie chart, bar chart, scattergraph, line graph and so on. The detailed operation of the graphing facilities varies from one spreadsheet package to another so they are not explained here.

Nugget: wizards

Readers of a certain age may remember *Wizzard* as a UK Midlands rock group with lead singer Roy (I wish it could be Christmas every day) Wood. A wizard (one 'z') on a computer application is a very useful tool that takes you, step-by-step, through a series of commands that would be hard to explain in words and hard to follow as a user. There are several wizards included with most spreadsheet applications, of which one of the most useful is called 'Chart Wizard'. This leads you by the nose through the steps required to transform your data into a pie chart, bar chart, scatterplot, etc.

However, if by now you have greater confidence with using a spreadsheet, this is something that you might like to try for yourself. When you feel ready, return to Chapters 3 and 4 and see if you can reproduce the graphs displayed there. You could also return to Chapter 5, Summarizing data, and practise calculating means, medians, standard deviations and so on using the spreadsheet.

Comments on exercises

▶ **Exercise 9.3**

The required formula for cell B6 is:

= B5–B4

Your final spreadsheet table should look like this:

Milk	0.74
Bread	0.88
Eggs × 12	2.65
TOTAL	4.27
TENDERED	10
CHANGE	5.73

▶ **Exercise 9.5**

Let's start by looking at the calculation required for the first child in the list, Bertie. In order to calculate the percentage change in

Bertie's score, you must first find the actual change, which is $50 - 38 = 12$, i.e. an increase of 12. To convert this to a percentage, this figure is divided by the 'before' score, 38, and the answer multiplied by 100. Expressed as a single calculation, this is:

$$\frac{50 - 38}{38} \times 100$$

On a spreadsheet, the corresponding formula (keyed into cell D2) is:

$$= (C2–B2)/B2 * 100$$

Note the use of brackets in this formula; these brackets are needed to make the division work correctly.

After 'filling down', the completed table should look like this:

	A	B	C	D
1	**PUPILS**	**Score before**	**Score after**	**Percentage change (%)**
2	Bertie	38	50	31.57894737
3	Moya	61	66	8.196721311
4	Erskine	18	59	227.7777778
5	Liam	34	30	-11.76470588
6	Patsy	44	60	36.36363636

Note that the numbers in Column D have been displayed with overly long decimal fraction answers. In this example, the teachers would probably be happy to have these results given to the nearest whole number. Make this change by selecting cells D2 to D6 and reformatting the cells to display to 0 decimal places. Depending on which spreadsheet package you are using, you can probably do this from the Format menu with Format —> Cells —> Number. This should give the following result.

	A	B	C	D
1	**PUPILS**	**Score before**	**Score after**	**Percentage change (%)**
2	Bertie	38	50	32
3	Moya	61	66	8
4	Erskine	18	59	228
5	Liam	34	30	-12
6	Patsy	44	60	36

You can now see at a glance that the big improver was Erskine with a gain of 228% and the pupil who benefited least was Liam, whose estimation skills actually showed a decrease of 12%.

Key ideas

This chapter looked at one of the most powerful computer applications that most people are ever likely to use: the spreadsheet.

Here are the main spreadsheet issues covered:

▶ The sort of information normally entered into a spreadsheet cell will be one of three basic types: numbers, text or formulas.

▶ Spreadsheets are particularly useful for carrying out repeated calculations – simply perform the first calculation and a 'fill down' (or 'fill right') command will complete all the other calculations automatically.

▶ What you see on a spreadsheet screen is only a window on a much larger grid, consisting of hundreds of rows and columns.

▶ In order to enter a formula, start the command with =.

▶ You have been introduced to a range of commands for calculating: the four operations (+, −, × and ÷), mean, median, minimum value, maximum value, range, lower quartile, upper quartile, inter-quartile range and standard deviation.

There are many others!

To find out more via a series of videos, please download our free app, Teach Yourself Library, *from the App Store or Google Play.*

10

Reading tables of data

In this chapter you will learn:

▶ *how to interpret tables of data*
▶ *how to reorganize information so as to make its main features stand out more clearly.*

For many people, there is something about a table of figures that seems to make their head swim. Indeed, as one reads a page of text that contains a table of data, there is a tendency to blank the table section out; the eye almost jumps over it as if it weren't there. It is certainly true that tables can look very uninviting and incomprehensible.

The aim of this chapter is to help you to extract what information is available in a table and also to become more proficient at laying out tables of your own data so that the key features are communicated effectively. This will require a certain attitude of mind about what job it is that you think a table of figures is designed to do. It is not simply there as a way of storing information; these days we have computer databases which will fulfil that role much more effectively. The purpose of a table, on paper, is to communicate information effectively to the person who is reading it. If it fails to do this, then it is a waste of space on the page.

Although the main thrust of this chapter deals with interpreting tables on a printed page, you will need to recognize that, increasingly, this sort of data handling is done on a computer using an application known as a spreadsheet. A description of spreadsheets, and how to use them, was given in Chapter 9.

Nugget: lamp-posts

It has been said that people sometimes tend to use tables of data like a drunk uses a lamp-post – more for support than for illumination!

Turning the tables

In this section you will look at the way in which information is often set out in a table and how, with a little common sense, it might be done better. We start with a set of data in Table 10.1, which is not well laid out. It shows the total population and urban population of certain European countries and other selected regions.

Exercise 10.1 What is it telling me?

Look at Table 10.1 and think about the following questions.

a Do you feel that the layout of the table could be improved? If so, in what way?

b What sort of information does the table reveal about urban population in these European countries and other selected regions?

c What additional information might you need or what further calculations might you wish to do in order to squeeze more useful information from it?

[Comments follow]

Table 10.1 Population and urban population of selected European countries and other selected regions (m means millions)

Country/region	Population (m) 2015	Urban population (m) 2015
Africa	1123	414.4
Australia	22.8	20.4
Austria	8.7	5.7
Belgium	11.3	11.1
Canada	35.1	28.7
China	1367	760.1
Denmark	5.6	4.9
Europe	742.5	526.4
Finland	5.5	4.6
France	66.6	52.9
Germany	81.9	61.7
Greece	10.8	8.4
Ireland	4.9	3.1
Italy	61.9	42.7
Luxembourg	0.6	0.5
Netherlands	16.9	15.3
Portugal	10.8	6.9
Spain	48.1	38.3
Sweden	9.8	8.4
UK	64.1	52.9
USA	321.4	261.3

Sources: The World Factbook (CIA), the World Bank and Wikipedia

A confusion here is that the countries and regions have been presented in alphabetical order, with the result that the European countries are mixed up with countries and regions from elsewhere. Another criticism of the layout is the rather untidy way that the figures have been written down: the data for Africa and China have been presented as whole numbers while others are given to one decimal place of accuracy. This causes two problems. First, it creates an inconsistency between the degree of accuracy to which the data from each country or region have been stated: the statement that 'the population of Africa in 2015 was 1123 million' is rather less precise than saying it was, say 1123.0 million. The second problem is that this inconsistency has the effect of throwing the numbers out of alignment, thereby making it awkward to run your eye down a column of figures and compare countries and regions. But we shall tackle these problems of layout later. Let's start by trying to make sure that the information being presented is the most useful that is available.

You may have felt that, as it stands, Table 10.1 doesn't tell you much. China, Africa and Europe have the largest urban populations, but this is hardly surprising since these regions have also the largest overall populations. Also, the information, even at the time of writing, is out of date. Having said that, there is always a time lag between the collection and the organization of data and an even longer delay until their publication, so *it is always the case that data are out of date!*

A more sensible way of comparing urban populations across these very different regions is to concentrate not on the absolute size of the urban population but instead to find the percentage of the population who live in urban areas. Let us take as an example the data for the first region in the table, Africa. To calculate the percentage of the 2015 population that lived in urban areas, calculate:

$$\frac{414.4}{1123} \times 100$$

This gives an answer of $36.901\ 157\ 614$.

Since the original data were given to one decimal place of accuracy, it makes sense to round this result to one decimal place, giving 36.9%.

We now need to perform the same calculation for each country or region. Phew! Although it sounds a daunting prospect, rest assured that I have already entered the data onto a spreadsheet and this sort of calculation is just what spreadsheets do best – in this case, dividing a column of figures by the corresponding figures in another column, and multiplying each result by 100. I have also ensured that each result has been corrected to the same number of decimal places – I returned to the original source and amended the data for China to one decimal place. As mentioned earlier, not only does this ensure that the data are internally consistent but also the columns of figures are now properly aligned, making for easier comparisons as you run your eye down a particular column. Table 10.2 shows the results of this operation.

Exercise 10.2 Spreadsheet (or calculator) check

If you have access to a spreadsheet, copy the data from the first three columns of Table 10.2 into columns A, B, C of your spreadsheet. Perform an appropriate calculation in Column D and 'fill down' to calculate the percentage urban data for 2015. Then check your results against the final column in Table 10.2 to see that you have done the calculation correctly.

If you don't have access to a spreadsheet, you can use your calculator to confirm your understanding of this calculation by checking some of the values in the final column of the table.

[Comments at the end of the chapter]

Table 10.2 Percentage urban population

Country/region	Population (m) 2015	Urban population (%) 2015
Africa	1123.0	37.0
Australia	22.8	89.4
Austria	8.7	66.0
Belgium	11.3	97.9
Canada	35.1	81.8

(Continued)

Country/region	Population (m) 2015	Urban population (%) 2015
China	1367.0	55.6
Denmark	5.6	87.7
Europe	742.5	70.9
Finland	5.5	84.2
France	66.6	79.5
Germany	81.9	75.3
Greece	10.8	78.0
Ireland	4.9	63.2
Italy	61.9	69.0
Luxembourg	0.6	90.2
Netherlands	16.9	90.5
Portugal	10.8	63.5
Spain	48.1	79.6
Sweden	9.8	85.9
UK	64.1	82.6
USA	321.4	81.6

Now that we have the data we want, the next stage is to try to improve the presentation and layout. Table 10.3 has been 'tweaked' by having the European countries separated from the rest.

Table 10.3 Populations of selected European countries and other selected countries/regions (2015)

Country/region	Population (m) 2015	Urban population (%) 2015
Austria	8.7	66.0
Belgium	11.3	97.9
Denmark	5.6	87.7
Finland	5.5	84.2
France	66.6	79.5
Germany	81.9	75.3
Greece	10.8	78.0
Ireland	4.9	63.2
Italy	61.9	69.0
Luxembourg	0.6	90.2
Netherlands	16.9	90.5
Portugal	10.8	63.5
Spain	48.1	79.6
Sweden	9.8	85.9
UK	64.1	82.6

Country/region	Population (m) 2015	Urban population (%) 2015
Europe	742.5	70.9
Canada	35.1	81.8
USA	321.4	81.6
Africa	1123.0	37.0
China	1367.0	55.6
Australia	22.8	89.4

And now it is time for the final reorganization. One or two further 'tweaks' are worth thinking about.

First, notice that the European countries have been presented in alphabetical order. While this is fine if you want to find the data for a particular country quickly, it is less than fine for seeing overall patterns in the data. A more useful strategy is to order the countries according to the size of the measure that you are most interested in. Unless there are other pressing reasons determining the order in which the rows or columns are to be presented, it is sensible to put the most important or interesting ones first. Let us assume, therefore, that it is the percentage urban population that is the particular factor of interest. I propose to sort this column so that it starts with the 'biggest first' on the grounds that the biggest values are usually the most interesting and the top part of the table is probably where most people look most carefully. Clearly, all the other columns have to follow suit so that the data for each country match up. On a spreadsheet, this is achieved using the **Sort** command.

When using **Sort** on a table like this, containing several linked columns, you need to be careful not to scramble the data. For example, if you were to sort one of these columns on its own, the data would no longer match up with the corresponding data in the other columns. Before sorting, you must therefore select the entire table (excluding the headers) and then carry out the sort by column B. This ensures that the rest of the table is also sorted in the same way. As you will see from looking at Table 10.4, I sorted the European countries and then the other regions separately.

Second, it is now time to reassess whether the one decimal place accuracy is the most appropriate for our purposes. Remember that we originally settled on one decimal place because it could be justified in terms of the accuracy of the data from which the rates were calculated. But it might well be that this level of precision is actually too great for the main purpose for which this table has been constructed, namely to look for interesting patterns. At this stage you could round the data to the nearest whole number. Table 10.4, which will be my final version, shows the data reordered for ease of interpretation.

Nugget: precision or accuracy

Two words that are commonly confused are 'precision' and 'accuracy'. They crop up because we can never make perfect measurements. A set of measures or estimates show *precision* if they are all very close together. *Accuracy* is shown when, on average, they are close to the true value. Another way of expressing this is that accuracy is the degree of veracity while precision is the degree of reproducibility.

Finally, here are these ideas shown pictorially, based on the idea that a perfect measure is one that strikes the centre of the bullseye.

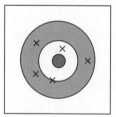

Accurate but not precise   Precise but not accurate   Both accurate and precise

Finally, with any set of data it is always worth bearing in mind what a typical or middle value looks like and how the others compare with it. Thus an extra row showing the median values for the European countries has been added directly below the final European entry. An italic typeface has been used to distinguish it from the other countries and the gap below it has been kept so that there is a clear separation between the two sets of countries/regions.

It is worth pointing out that I have presented this example in four separate and rather drawn-out stages. In practice, of course, several of these stages would be collapsed into one. A great virtue of performing the task on a spreadsheet is that you can try out many possible layouts very easily before settling on the one that is the most effective.

Table 10.4 Population and urban population of selected European countries and other selected countries/regions (2015): the final version

Country/region	Population (m) 2015	Urban population (m) 2015	Urban population (%) 2015
Belgium	11.3	11.1	98
Netherlands	16.9	15.3	91
Luxembourg	0.6	0.5	90
Denmark	5.6	4.9	88
Sweden	9.8	8.4	86
Finland	5.5	4.6	84
UK	64.1	52.9	83
Spain	48.1	38.3	80
France	66.6	52.9	80
Greece	10.8	8.4	78
Germany	81.9	61.7	75
Italy	61.9	42.7	69
Austria	8.7	5.7	66
Portugal	10.8	6.9	64
Ireland	4.9	3.1	63
Median European	*10.8*	*8.4*	*80*
Australia	22.8	20.4	89.4
Canada	35.1	28.7	81.8
USA	321.4	262.3	81.6
Europe	742.5	526.4	70.9
China	1367.0	760.1	55.6
Africa	1123.0	415.5	37.0

Now that the table has been reorganized to our satisfaction, let us return to the central question on which this exercise in data interpretation was based.

What are the main conclusions to be drawn from running an eye down the final column of Table 10.4? Perhaps one overall theme is that, in the developed world (Europe, US, Canada and Australia), levels of urbanization are consistently high. So it would seem that the wide-open spaces of Canada and Australia are clearly home to a relatively small number of their citizens and the majority of their citizens are concentrated in towns and cities.

The key contrast is with Africa and China, which still have largely rural economies. You may wonder what other economic indicators are linked to high levels of urbanization; it could be the case that, in general, wealthier countries also tend to be more urbanized, but it would require more data and further investigation to confirm this.

Your analysis has also revealed that the most highly urbanized parts of the European countries in the table are the Benelux countries: Luxembourg, Belgium and the Netherlands. At the other end of the scale are Portugal and Ireland, which still have relatively rural populations. You may be wondering why greater urbanization takes place in some countries rather than others, but these questions are not answerable from the data provided here.

To end this section, here are two additional handy hints concerning the layout and presentation of a table of data.

▶ For certain types of tables it may be useful to add together all the row totals and also add together the column totals. The sum of the row totals should equal the sum of the column totals and this can provide a check for internal consistency.

▶ Trying to read a table that contains too many rows or columns can be overwhelming, in which case it may be possible to collapse some of the rows or columns together to make a smaller, simpler table.

Reading the small print

If people's eyes tend to blank out tables of figures, you can be darned sure that they blank out the small writing that goes around them.

Alan Graham, 2017

Reading tables of data intelligently can be quite a creative exercise, drawing on the use of a range of detection skills in order to gain some better understanding of the story that lies behind the figures. So far the focus has been entirely on the main body of the table and how we might reorganize or recalculate the data that it contains. However, there are often many other vital – but less visible – clues around the edges. In particular, the title, source and footnotes may provide essential background information which may require you to qualify or refine any conclusions you make.

Finally, a few words about title and source. A good title should be both clear and concise. Sometimes it is hard to be both of these things at the same time, in which case be concise and put any additional qualifications or details in a footnote. A common area of confusion in titles dealing with British data is whether the data refer to the UK, to Great Britain or to England & Wales. These are all different land masses with different (though overlapping) populations and careful reading of the title should reveal which you are dealing with. The importance of giving a source is to enable the reader to check your data and perhaps to chase up further information about them.

Exercise 10.3 What does it all mean?

Look at Table 10.5.

a Which parts of the United Kingdom are included and which are excluded in these data?

b Why might the term 'household size' be ambiguous?

c Using a spreadsheet, confirm that the data on 'Average household size' in the final row of the table are consistent with the rest of the table.

[Comments at the end of the chapter]

Table 10.5 Households by size

Great Britain	Percentages				
	1971	1981	1991	2001	2011
One person	18	22	27	29	31
Two people	32	32	34	35	34
Three people	19	17	16	16	16
Four people	17	18	16	14	13
Five or more people	14	11	7	7	7
All households (= 100%) (millions)	18.6	20.2	22.4	23.8	26.4
Average household size (number of people)	2.9	2.7	2.5	2.4	2.3

Source: adapted from *Social Trends* 37, Table 2.1 and the Office for National Statistics

Nugget: very small print indeed

During a lecture in 1959, entitled 'There's Plenty of Room at the Bottom', the charismatic physicist Richard Feynman suggested that there were no physical barriers to 'nano' technology. At the end of his talk, he offered a $1000 prize to anyone who could rewrite a page from an ordinary book so that the text was shrunk by a factor of 25 000 (at this scale, the entire contents of the *Encyclopaedia Britannica* would fit onto the head of a pin). He had to wait 26 years before handing over his money to Stanford graduate student Tom Newman, who used electron beam lithography to engrave the opening page of Charles Dickens' *A Tale of Two Cities* in such tiny print that it could be read only with an electron microscope.

Comments on exercises

▶ **Exercise 10.2**

The spreadsheet formula for cell D2 was:

$$= C2/B2*100$$

▶ **Exercise 10.3**

a   Great Britain includes England, Scotland and Wales. The United Kingdom covers these three countries plus Northern Ireland, so Northern Ireland is excluded from the data in Table 10.5.

b   It isn't obvious exactly what types of social groupings this term 'household' covers. For example, would it include large communal houses such as boarding schools or homes for the aged? It is possible that some countries define 'households' differently from others, in which case international comparisons become rather dubious. For the purpose of this survey it is defined in terms of people living together in a domestic setting and sharing facilities.

c   A spreadsheet approach to checking the final row of data in the table is suggested below.

As an example, here is one way of checking the figure for 'Average household size of 2.4' for 2011. First, Column A needs to be replaced by numbers, 1, 2, 3, etc. A problem occurs with the final category of household size, 'Six or more people', which could include 6, 7, 8, ... people up to anything like 15 or 20. How can we take a representative number for this category? This problem was discussed in Chapter 5, Summarizing data, in the calculation of a weighted mean. This time, since the number of very large households is much lower than before, we shall adopt a smaller household size of 7 to represent this 'Six or more people' category.

Enter the following data into Columns A to C of a blank spreadsheet.

	A	B	C	D
1		**Size**	**2011**	
2	One person	1	31	
3	Two people	2	34	
4	Three people	3	16	
5	Four people	4	13	
6	Five or more people	5	7	

You can now use the spreadsheet to calculate the weighted mean of the data in Columns B and C, as follows. In cell D2 enter the following formula:

$$= B2*C2/100$$

Then fill down from D2 to D6. Click in D7 and enter the following formula:

$$= SUM(D2:D6)$$

This produces the weighted mean of the values in column B, weighted by the values in column C, of 2.34. Your final spreadsheet table should look like this.

	A	B	C	D
1		**Size**	**2011**	**Product/100**
2	One person	1	31	0.31
3	Two people	2	34	0.68
4	Three people	3	16	0.48
5	Four people	4	13	0.52
6	Five or more people	5	7	0.35
7			Weighted mean	2.34

The answer of 2.34 is consistent with the value of 2.3 shown in Table 10.5 (rounded to one decimal place).

Key ideas

This chapter looked at different aspects of the design of tables of data. Some useful strategies that were mentioned for tidying up and simplifying tables included the following:

▶ Remove unnecessary rows and columns (too much information is off-putting).

▶ Remove unnecessary digits (too much precision is off-putting).

▶ Re-order rows and columns where necessary in order to help the main features stand out.

▶ Append row and column totals and/or averages. The totals can be used as a check for internal consistency as well as giving an overview.

▶ Raw 'absolute' data may be unsuitable for making comparisons, in which case it may be possible to change them into a 'relative' form, such as rates or percentages.

▶ Round the data to as few figures as you dare. Remember that the central purpose of the table is less to store data than to reveal patterns that they might contain.

▶ The table may not provide all the information you need, in which case you may need to check the original source or chase up a further source.

▶ Read the small print carefully, especially the title, source and footnotes.

To find out more via a series of videos, please download our free app, Teach Yourself Library, *from the App Store or Google Play.*

11

Regression: describing relationships between things

In this chapter you will learn:

- ▶ *how to graph paired data*
- ▶ *how to show the overall trend using a 'best fit' line*
- ▶ *how to make predictions from paired data*
- ▶ *how to calculate a regression line.*

You will find a calculator helpful for this chapter.

Paired data

Data collection generally takes place from a sample of people or items. Depending on the purpose of the exercise, we often take just *one single* measure from each person or item in the sample. For example, a nurse might wish to take a group of patients' temperatures, or a health inspector might measure the level of pollution at various locations along a river. These sorts of sample will consist of a set of single measurements which should help to provide a picture of the variable in question. So, the nurse could use the temperature data to gain some insight into the average temperatures of patients in hospital and how widely they are spread. Similarly, the health inspector could use the pollution data to establish the typical levels of pollution present in the river, and also to see to what extent the levels of pollution vary from place to place.

Sometimes, however, data can be collected *in pairs*. Paired data allow us to explore something quite different from what is indicated by the examples above. Using paired data, we can look at the possible *relationship* between the two things being measured. To take a simple example, look at the two graphs, Figure 11.1, which show the typical amounts of sunshine and rainfall over the year in the Lake District.

Exercise 11.1 Hot dry summers?

a How would you describe the pattern of sunshine over the year?

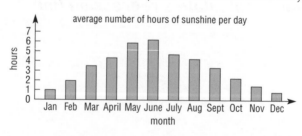

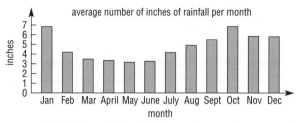

Figure 11.1 Weather in the Lake District

Sources: various.

b How would you describe the pattern of rainfall over the year?

c Putting (a) and (b) together, is there any pattern in the relationship between the amount of sunshine and the amount of rainfall over the year?

[Comments below]

a Generally, the sunniest months are in the middle of the year, around May and June, while the months at the beginning and end of the year have the least sunshine.

b Rainfall is least in the middle of the year, around May and June, while the months at the beginning and end of the year have most rainfall.

c It seems clear from (a) and (b) that there is a relationship between rainfall and sunshine. In general, the months which have the most sunshine tend to have the least rainfall, and vice versa. The pattern in this relationship can be spelled out more clearly when drawn on a scattergraph.

If you are in any doubt about how to plot the points, you can check your scattergraph with the one given in Figure 3.14. As you should see from your scattergraph, the pattern of points is fairly clear. The points seem to lie roughly around a straight line which slopes downwards to the right. This bears out the conclusions given to Exercise 11.1 part (c), namely that high sunshine tends to be associated with low rainfall. The main ideas of this chapter and the one which follows it are all about *interpreting more precisely the sort of patterns that can be found on a scattergraph.* A key point to remember, however, is that these ideas involve investigating relationships between two things and therefore throughout this chapter we will be dealing with *paired data.*

Exercise 11.2 Graphing paired data

Using the graphs in Figure 11.1, copy and complete Table 11.1. Then plot these pairs of data on a scattergraph such as the one in Figure 11.2. (If you feel it necessary, reread the section on graphs in Chapter 2.)

Table 11.1 Data from the graphs in Figure 11.1

Month	Average hours of sunshine (estimated)	Rainfall (estimated)
January	1.0	6.6
February		
March		
April		
May		
June		
July		
August		
September		
October		
November		
December		

[Comments at the end of the chapter]

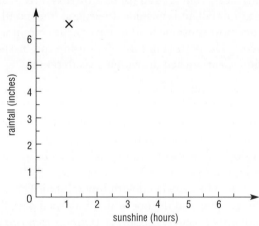

Figure 11.2 Scattergraph showing the relationship between the amount of sunshine and rainfall over a typical year

Source: data from Table 11.1.

The 'best-fit' line

The word used to describe the overall shape of points plotted on a scattergraph is the *trend*. It seems clear from the previous example that the overall trend in Figure 11.2 is a downward-sloping one. However, it is possible to be a bit more precise about this pattern. Imagine drawing a straight line through the middle of these points.

Nugget: fitting a line

By the way, the phrase 'drawing a line through a set of points' is open to misunderstanding. Students who enjoyed creating dot-to-dot drawings in their childhood sometimes interpret this as joining up the dots on the scattergraph. This is not what is meant here! The procedure of fitting a (straight) line through a set of points is a bit like trying to place a knitting needle on top of the scattergraph so that its position and direction is the best summary of where the points lie. As such, it may not actually pass through many (or indeed any) of the points.

Finding the equation of such a line would allow us to move from vague, wordy descriptions of the trend to a more precise mathematical description of the relationship in question. Once the line has been defined in this more formal way, it becomes possible to make predictions about where we think other points might lie. Incidentally, this involves using a few basic mathematical ideas about graphs and coordinates and these are explained in Chapter 2.

The best-fit line or *regression line*, as it is sometimes called, is the best line which can be made to fit the points which it passes through. However, a straight line may be rather too crude a description of the trend. In the world of economic and industrial forecasting, for example, curves are commonly used and these provide a necessary degree of subtlety and accuracy. For instance, economic data are often subject to seasonal fluctuations (people tend to spend more money on household goods around December and January and less in the summer).

Fitting straight lines to these sorts of data would be to ignore this seasonal factor and result in highly inaccurate predictions. But, for the purposes of this chapter, we will look only at methods using straight lines. This is called *linear* (i.e. straight line) regression.

Exercise 11.3 By eye

a Using a ruler and pencil, draw by eye the best-fit straight line onto your scattergraph from Exercise 11.2.

b Think about how you might describe this line mathematically.

[Comments below and at the end of the chapter]

In order to define a straight line exactly, two pieces of information are required. By convention, these are usually the *slope* of the line, and the point on the graph where it crosses the vertical axis (known as the *intercept*). The slope can be either a positive or a negative number. Lines which slope from bottom left to top right will have a positive slope, i.e. the value of Y will increase as the value of X increases. Lines which have a negative slope run from top left to bottom right. These ideas are explained in Chapter 2 and, if you are unfamiliar with them or find them confusing, you should go back and reread the section 'Graphs' now.

In this example, rainfall is on the vertical axis and sunshine on the horizontal axis. The conventional labels for these axes are, respectively, Y and X. In general, if we are seeking to find the linear regression line, the equation of the straight line required will be of the form:

$$Y = a + bX$$

where Y represents rainfall (measured in inches per month)
$\quad\quad\,\, X$ represents sunshine (measured in hours per day)

and *a* and *b* are the two pieces of information required to define the line:

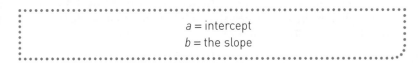

$$a = \text{intercept}$$
$$b = \text{the slope}$$

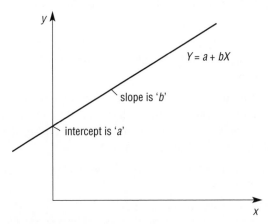

$$Y = a + bX$$

slope is '*b*'

intercept is '*a*'

Figure 11.3 The regression coefficients

Incidentally, as was explained in Chapter 2, there are several possible ways of setting up this equation. First, you might prefer to use labels which are easier to remember than X and Y; R and S, for instance. If so, the regression equation would look like this[1]:

$$R = a + bS$$

a and *b* are known as the regression coefficients.

---

[1] You may be more familiar with straight line equations which have the form '$Y = mX + c$' than the one given here. However, in statistics, the form '$Y = a + bX$' is the more common and, if you have a calculator which finds regression lines, the letters '*a*' and '*b*' are likely to be the labels used on the relevant keys.

Now that you have drawn in the best-fit line by eye (Exercise 11.3, part a) the next stage is to find its equation. The intercept, which is the '*a*' value in the equation, is quite easy to read off the graph, but the slope (the value of '*b*' in the equation) is less straightforward. Try it yourself first in Exercise 11.4.

Exercise 11.4 Finding the regression coefficients, *a* and *b*, from the graph

a Read off the intercept where the line passes through the vertical axis. This is the '*a*' value in the equation of the best-fit line.

b Remember that the slope of a line is the amount it increases for each unit it travels horizontally from left to right (a decrease is measured as a negative quantity). From the graph estimate the slope, i.e. the '*b*' value in the best-fit line.

c Using these estimates for '*a*' and '*b*', write down the equation of the best-fit line.

[Comments at the end of the chapter]

In practice, best-fit equations are rarely found in this way but Exercise 11.4 was included to give you an intuitive feeling for what is involved in calculating regression coefficients. There are several other methods for finding the values of *a* and *b*, the simplest of which is to feed the raw data directly into a suitable calculator (or computer) and obtain them at the press of a button. When all the data have been entered, you then simply press the keys marked '*a*' and '*b*' to discover the values for the intercept and slope of the best-fit line. If you do have a suitable calculator which handles linear regression, spend a few minutes now entering the paired data given in Table 11.6. If you have never done this before, be prepared to spend some time reading through your calculator manual.

Alternatively, if you do not have a calculator that will calculate the regression coefficients, there is a formula which allows you to calculate them directly. This is included later in the chapter, so turn to this section now if you don't have a suitable calculator or wish to see how to calculate the regression coefficients using the formula. If you aren't interested in the mathematical aspects of calculating the values of *a* and *b*, you might prefer to omit this section. However, it would still be helpful to have a look at Figure 11.9, which shows the best-fit line and how the values of *a* and *b* show up on the graph.

Nugget: don't mix up your *a* and your *b*!

If you use a calculator or computer to find the values of the regression coefficients, *a* and *b*, be careful that you are clear which is which. The problem here is that different conventions operate in different countries as to whether the regression equation should be described as $y = a + bx$ or $y = ax + b$. In the first case, the *a* refers to the value of the intercept and *b* is the slope; in the second case, the reverse is true. Which of these two models has been used should be clearly stated somewhere on the screen, so make sure you check this out.

In this example the slope had a negative value (-0.6), which was reflected in a downward-sloping line. This type of trend suggests a *negative relationship* between sunshine and rainfall (i.e. as one increases the other tends to decrease). Where the slope is positive, we say that there is a positive relationship between the two things under consideration (i.e. as one increases the other also tends to increase).

At the level at which this book is aimed, it is enough to know what the regression coefficients mean (i.e. that they give the slope and the intercept of the regression line) and to be able to calculate them, preferably using a suitable calculator. However, it is also worth knowing that the best-fit line through a set of points has one important property, namely that, if you take the vertical distances of each point from this line, square them and add the squares together, the result is smaller than for any other straight line you might have chosen. In fact, the formula given in the Comments on exercises derives from this so-called 'least squares' property. For this reason, then, this line of best-fit is also known as the least-squares line. In other words, the following terms all mean more or less the same thing:

▶ best-fit line

▶ least-squares line

▶ linear regression line.

The next exercise will give you the chance to calculate a best-fit line for yourself and also to give some thought as to how it might be interpreted.

Exercise 11.5 Hours of study

Table 11.2 gives the number of hours of study that a student, Marsha, spent on each of six modules, along with the test score she got for each.

Table 11.2 Hours of study and scores

Module	Hours of study	Score
A	5	66
B	12	72
C	3	44
D	1	54
E	16	86
F	8	70

a Plot the data on a scattergraph, such as the one shown in Figure 11.4. Estimate by eye the slope and intercept from the graph.

b Using a spreadsheet, a statistical calculator, or by the long method given later in the chapter, find the equation of the best-fit line. Draw it on your graph and check that it actually looks like a line of best fit.

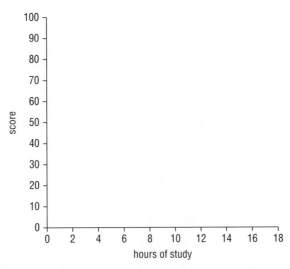

Figure 11.4 Unfinished scattergraph for the data in Table 11.2

  c Does the best-fit line have a positive or a negative slope? What does this mean in terms of these two variables?

  [Comments at the end of the chapter]

Making predictions

By now you should have mastered the technicalities of calculating the regression coefficients and drawing the best-fit line through a scatter of points. Essentially, this best-fit line is a mathematical summary of what you think might be the relationship between the two variables in question. However, what is more interesting than calculating it is to use and interpret it. What we can now do is to use the line to make predictions about other pairs of values, perhaps by extending the line or by making guesses about points on the line where data are absent. However, it is worth stressing that these really are only guesses which carry all sorts of assumptions that may not hold in practice. Now do Exercise 11.6.

Exercise 11.6 Racing pulses?

Pulse rates and ages have been recorded for 14 people and are shown in Table 11.3.

a Plot the data onto a scattergraph such as the one in Figure 11.5. What does the pattern of points suggest to you about the relationship between someone's age and their pulse rate?

b Either by eye, or by calculating the regression coefficients, draw the linear regression line (i.e. the best-fit line) through your points.

c Use the regression line to make the following predictions:

▶ Mia is 60 years old. What pulse rate would you expect her to have?

▶ Mayerling is 2 years old. What pulse rate would you expect her to have?

▶ Tom has a pulse rate of 90 beats per minute. How old do you think he is?

d What assumptions have you made in carrying out these predictions?

[Comments at the end of the chapter]

As you will have read in the comments to this exercise, there are considerable dangers in using a regression line to make estimates of other values. In general, these dangers are least when estimates are made within the range of values covered by the sample – for example in the case of the prediction made about Mia's pulse rate. Predictions of this sort, which lie within the range of the sample, are called *interpolation* (literally 'within the points'). Of course, even interpolated estimates may be extremely inaccurate for a number of reasons; the regression line is only an average, after all, and merely gives rough predictions. Thus, it would be silly to quote Mia's predicted pulse rate, say, to two decimal places, as 72.04. A prediction of 'about 72' or even 'about 70' would be more realistic.

Table 11.3 The ages and pulse rates of a sample of 14 adults

Name	Age (years)	Pulse rate (beats per minute)
Ellen	33	62
Paul	50	66
Ramesh	45	77
Karl	65	78
Donna	57	78
Eliza	70	70
Siobhan	42	62
Joseph	75	76
Joanne	57	67
Debbie	35	70
Indira	44	68
Sean	63	70
Beena	50	70
Tara	55	72

Source: personal survey.

The dangers of inaccuracy increase when making predictions which involve having to extend the line beyond the range of values in the sample. Such predictions are known as *extrapolation* (literally going 'beyond the points'), and the estimates made about Tom and Mayerling are examples of extrapolating or extending the regression line. In this example, both these predictions are almost certainly wildly inaccurate. In the case of Tom, most people aged 138 tend to have a pulse rate much closer to zero beats per minute than 90! What is more likely is that Tom had just completed a bout of intensive exercise prior to having had his pulse taken. Predicting Mayerling's pulse rate involved extrapolating in the other direction. As before, we have assumed that the pattern which seems to be true for the sample of 14 older people also holds true for children. Unfortunately, it is just not the case. In fact, for children, the relationship between age and pulse rate tends to be a negative, rather than a positive, one as the data in Table 11.4 and Figure 11.6 illustrate.

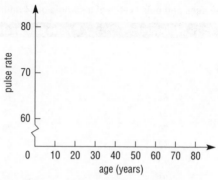

Figure 11.5 Scattergraph showing the data from Table 11.3

Table 11.4 The ages and pulse rates of a sample of ten young people

Name	Age (years)	Pulse rate (beats per minute)
Irena	10	86
Nadia	3	107
Luke	4	98
Jon	6	105
Carrie	8	90
Theo	1	115
Jagieet	16	69
Ira	21	81
Hannah	5	100
Alex	2	105

Source: personal survey.

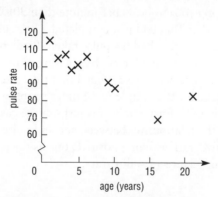

Figure 11.6 Scattergraph showing the data from Table 11.4

Seeing these additional data may now encourage you to reconsider the case of Tom. On the basis of the second regression line you have drawn, someone with a *regular* pulse rate of 90 beats per minute might be expected to be 10 or 11 years of age.

Taken separately, the pulse rate data from these two separate samples seemed to indicate that linear regression was a valid way of summarizing the trends in each case. However, when we start to consider the two samples taken together – that is to look at the relationship between pulse rate and age over the entire age range – we begin to question whether linear regression is a sensible way of describing what we see. If the two samples are collapsed together and the entire 24 points are plotted onto the same graph, a very different picture emerges (see Figure 11.7).

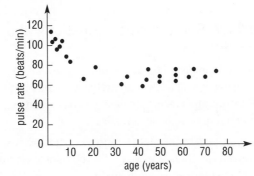

Figure 11.7 Scattergraph showing the data from Tables 11.3 and 11.4 together

There are methods for finding a non-linear curve which fits this pattern – a sort of 'curve of best-fit' – but these are beyond the scope of this book.

Nugget: non-linear regression

If you should happen to use a graphical calculator, such as the excellent Texas Instruments TI–84 Plus, you will find that a variety of different regression models are available. For example, you would select option 4 for linear regression. If the arrangement of points suggest an underlying cubic pattern, choose option 6 for cubic regression, and so on.

```
EDIT CALO TESTS
1:1–Var Stats
2:2–Var Stats
3:Med–Med
4:LinReg(ax + b)
5:QuadReg
6:CubicReg
7↓QuartReg
```

Now to end this section on regression lines, Exercise 11.7 allows you to think more deeply about aspects of regression which may have concerned you.

Exercise 11.7 Some further questions about the line of best fit

a A dictionary definition of the word 'regression' will be something like 'going back to an earlier or more primitive form'. How does this tie in with what you have read in this chapter about regression in statistics?

b Suppose you wish to calculate a regression line where one of the chosen variables was 'year'. Do you think it would be necessary to enter each year in full (e.g. 2001, 2002, 2003, ...) or could you enter them as 1, 2, 3, etc.?

c If the scales on either of the two axes are changed, will this alter the slope of the best-fit line?

d So far, the labels for the axes of each of the scattergraphs have been chosen for you, so you have not had to choose which way round they should go. Do you think it matters which way round they are?

e What are the most common types of data for which regression lines are used?

f How good do you think the fit has to be for the best-fit line to be meaningful?

[Comments follow]

a There is a clear link between the statistical meaning of 'regression' and its everyday use. The regression line can be thought of as a simplified (or 'primitive') summary of the relationship in question around which the points are scattered.

b It is perfectly legitimate to enter these values as 1, 2, 3, etc. This is equivalent to subtracting 2000 from the original numbers and doesn't alter the shape of the regression line but merely moves it to a different position on the graph. Thus, as long as the year axis is scaled '1, '2, '3, etc., it will look exactly the same.

c Changing the scales of either axis may alter the slope in terms of how steep it *looks* on the graph but, since the actual value for '*b*' remains unchanged, the effect is purely visual. The important general point to come out of this is that you shouldn't pay too much attention to the way the slope of the graph looks, as this is greatly affected by the scales used on the axes and whether or not there is a 'break' on the vertical axis. This and other related questions were discussed more fully in Chapter 6, Lies and statistics.

d The choice of which variable should go on which axis is usually *not* arbitrary. By convention, we try to choose the variable which is clearly not dependent on the other and place it on the horizontal axis. For example, it seems reasonable to assume that a person's age is independent of their pulse rate, rather than the other way around. So, in Figures 11.5 and 11.6, the variable 'Age' has been placed along the horizontal axis. However with some scattergraphs the choice may be more arbitrary, as with rainfall and sunshine, for example. It isn't obvious here which of these variables can be better thought of as being independent of the other. This decision and the whole question of dependence will be discussed in more detail in the next chapter.

e The most common types of data where regression is used are those where the variable 'time' is drawn on the horizontal axis – known as 'time series' graphs, for obvious reasons. Extrapolations made from time series graphs allow us to

make backward or forward projections, which are interesting
if we wish to make historical guesses about the past, or
forecasts about such things as the population and the
economy in the future.

f   Clearly, if the scattergraph shows up a very clear linear
pattern, then the best-fit line will be an extremely
accurate summary of the relationship with all the points
lying quite close to the best-fit line. In this situation
predictions can be made with a good degree of confidence.
However, if the points on the original scattergraph had no
clear pattern, with a very wide scatter, then the line will
have little predictive power. Just how confident you can be
in the predictions that you make from the regression line
will be discussed as the central theme of the next chapter.

Calculating the regression coefficients using the formula

An accurate method for finding the regression coefficients is to
perform a rather complicated calculation, using the formulas
given below. Later in this section you will be asked to find the
regression coefficients for the sunshine/rainfall data given in
Table 11.6.

To find the slope, b, you first need to calculate the following
intermediate values:

n $\rightarrow$ this is the number of pairs of values (in this example,
$n = 12$)

$\Sigma XY \rightarrow$ the sum of the 12 XY products

$\Sigma X \rightarrow$ the sum of the X values

$\Sigma Y \rightarrow$ the sum of the Y values

$\Sigma X^2 \rightarrow$ the sum of the squares of the X values

These intermediate values are then substituted into the following exciting-looking formula (which will not be justified but simply stated here).

$$b = \frac{n\Sigma XY - \Sigma X \Sigma Y}{n\Sigma X^2 - (\Sigma X)^2}$$

Having calculated the value for b, finding a follows more easily. The mid-point of all the points on the scattergraph is denoted by the coordinates $(\overline{X}, \overline{Y})$. (*Note*: the sigma ($\Sigma$) and bar $(\overline{X}, \overline{Y})$ notations used above are explained in Chapter 2.)

An important property of the best-fit line is that it passes through this mid-point. That being so, we can substitute these coordinates, $(\overline{X}, \overline{Y})$, into the equation of the line $Y = a + bX$. This gives:

$$\overline{Y} = a + b\overline{X}$$

Rearranging this equation produces the following:

$$a = \overline{Y} - b\overline{X}$$

from which we can find the value of a.

The example below uses this method to calculate the regression coefficients a and b based on the rainfall/sunshine data. In order to simplify the calculation, we shall call inches of rainfall Y and hours of sunshine X.

First, let us find b. From the formula for b shown above, it will be necessary to find the following intermediate values: ΣXY, ΣX, ΣY and ΣX^2.

Knowing what we need to find, the next stage is to set up a table of values so that these intermediate values can be calculated systematically. A suitable table might look something like Table 11.5.

Table 11.5 Calculating the regression coefficients

Month	Hours of sunshine (X)	Inches of rainfall (Y)	X²	XY
January	1.0	6.6	$(1.0)^2 = 1.00$	$1.0 \times 6.6 = 6.60$
February	1.8	4.1	$(1.8)^2 = 3.24$	$1.8 \times 4.1 = 7.38$
March	3.3	3.3	10.89	$3.3 \times 3.3 = 10.89$
April	4.3	3.2	18.49	$4.3 \times 3.2 = 13.76$
May	5.8	3.0	33.64	$5.8 \times 3.0 = 17.40$
June	6.1	3.1	37.21	$6.1 \times 3.1 = 18.91$
July	4.7	4.3	22.09	$4.7 \times 4.3 = 20.21$
August	4.3	5.0	18.49	$4.3 \times 5.0 = 21.50$
September	3.4	5.6	11.56	$3.4 \times 5.6 = 19.04$
October	2.3	6.8	5.29	$2.3 \times 6.8 = 15.64$
November	1.5	6.0	2.25	$1.5 \times 6.0 = 9.00$
December	0.9	5.9	0.81	$0.9 \times 5.9 = 5.32$
TOTALS	$\Sigma X = 39.4$	$\Sigma Y = 56.9$	$\Sigma X^2 = 164.96$	$\Sigma XY = 165.64$
$n = 12$, so:	$\bar{X} = \dfrac{39.4}{12}$	$\bar{y} = \dfrac{56.9}{12}$		

Note that when using a calculator it is not necessary to fill in each cell in the table above since it is only the totals we are interested in. Here is the formula for b again:

$$b = \frac{n\Sigma XY - \Sigma X \Sigma Y}{n\Sigma X^2 - \left(\Sigma X\right)^2}$$

By substitution from the table, we get:

$$b = \frac{12 \times 165.64 - 39.4 \times 56.9}{12 \times 164.96 - (39.4)^2}$$

$$= \frac{-254.18}{427.16}$$

$$= -0.595 \text{ (to three decimal places)}$$

This is the gradient (i.e. slope) of the regression equation.

Having found the value of b, we use it to find a.

$$\left(\text{Notice that } \overline{Y} = \frac{\Sigma Y}{n} \text{ and } \overline{X} = \frac{\Sigma X}{n}\right)$$

$$\begin{aligned}
a &= \overline{Y} - b\overline{X} \\
&= \frac{\Sigma Y}{n} - b \times \frac{\Sigma X}{n} \\
&= \frac{56.9}{12} - b \times \frac{39.4}{12} \\
&= 4.742 - (\text{-}0.595) \times 3.283 \\
&= 6.695 \text{ (to three decimal places)}
\end{aligned}$$

This is the intercept of the regression equation.

So, the regression equation is:

$$Y = 6.695 - 0.595X$$

Finally, it is worth reminding yourself what the variables X and Y were used to represent.

X = Average number of hours of sunshine per day
Y = Average number of inches of rainfall per month

So, 'Average inches of rainfall' = $6.695 - 0.595 \times$ 'Average hours of sunshine'.

What this means is that, if you know that the average number of hours of sunshine for a particular month in the Lake District was, say, 5 hours, then the predicted rainfall for that month could be estimated using the formula:

Hence, average inches of rain = $6.695 - 0.595 \times 5$
$= 6.695 - 2.975$
$= 3.72$

Comments on exercises

▶ **Exercise 11.2**

Reading from the two bar charts gives the estimates shown in Table 11.6.

The 12 points are plotted as shown in Figure 11.8.

Table 11.6 Sunshine and rainfall data estimated from Figure 11.1

Month	Sunshine (hours)	Rainfall (inches)
Jan	1.0	6.6
Feb	1.8	4.1
Mar	3.3	3.3
April	4.3	3.2
May	5.8	3.0
June	6.1	3.1
July	4.7	4.3
Aug	4.3	5.0
Sept	3.4	5.6
Oct	2.3	6.8
Nov	1.5	6.0
Dec	0.9	5.9

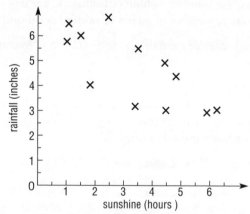

Figure 11.8 Scattergraph showing the data from Table 11.6

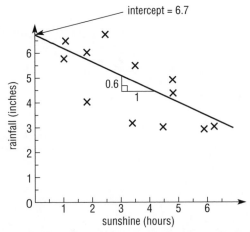

Figure 11.9 Best-fit line showing the relationship between rainfall and sunshine

a From Figure 11.9, the best-fit line cuts the Y axis at around 6.7 inches of rainfall. So, $a = 6.7$.

b Figure 11.9 has focused on a small section of the best-fit line which is one unit wide. Over that interval of one unit in the X direction, the graph has fallen by around 0.6 units in the Y direction. Because this is a *fall*, it must be recorded as a *negative slope*. So, $b = -0.6$. (A more accurate approach might be to choose a wider interval of, say, 4 units in the X direction, measure the fall in the Y direction over this interval and then divide by 4 to get the average fall per unit interval.)

c Substituting the values for a and b into the general equation, $Y = a + bX$, we get: $Y = 6.7 - 0.6X$.

► Exercise 11.5

a The scattergraph (and regression line) will look like Figure 11.10.

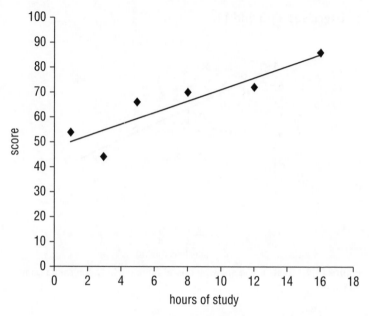

Figure 11.10 Scattergraph showing the relationship between hours of study and test score

b The regression line, or 'line of best fit', has the equation:

$Y = 47.9 + 2.3X$, where Y refers to 'score' and X to 'hours of study'

As you can see, the regression line has been drawn on the graph and clearly does seem to fit the points well. Note that the general regression equation is written as $Y = a + bX$, so you will be able to work out the equation for this particular example when you have calculated the regression coefficients a and b. Table 11.7 illustrates the formula method for calculating the regression coefficients for the 'hours of study/test score' data.

Table 11.7 Calculating the regression coefficients for the hours of study/
test score data

Hours of study (X)	Test score (Y)	X²	Y²	XY
5	66	25	4356	330
12	72	144	5184	864
3	44	9	1936	132
1	54	1	2916	54
16	86	256	7396	1376
8	70	64	4900	560
TOTALS	$\sum X = 45$	$\sum Y = 392$	$\sum X^2 = 499$ $\sum Y^2 = 26\,688$ $\sum XY = 3316$	

$n = 6$, so: $\overline{X} = \dfrac{45}{6} = 7.5$ $\overline{Y} = \dfrac{392}{6} = 65.333$

Source: personal survey.

Here again is the formula for b, the gradient of the regression line:

$$b = \frac{n\sum XY - \sum X \sum Y}{n\sum X^2 - \left(\sum X\right)^2}$$

By substitution from the table into the formula, we get:

$$b = \frac{6 \times 3316 - 45 \times 392}{6 \times 499 - 45}$$

$$= \frac{2256}{696}$$

$$= 2.328$$

$$= 2.328 \text{ (to three decimal places)}$$

Remember that this value, b, is the gradient of the regression equation.

Having found the value of b, we use it to find the other regression coefficient, a, which is the intercept on the graph.

Since the mean X and Y values (written as $\overline{X}$ and $\overline{Y}$) lie on the regression line, we can write:

$$\overline{Y} = a + b\overline{X}$$

Rearranging, this gives:

$$a = \overline{Y} - b\overline{X}$$

$$= 65.333 - 2.382 \times 7.5$$

$$= 47.9 \text{ (to one decimal place)}$$

The values for a and b must now be substituted into the general regression equation, $Y = a + bX$. So, the particular regression equation is:

$$Y = 47.9 + 2.3X$$

Where:

$$X = \text{Number of hours of study}$$

and

$$Y = \text{Test score}$$

c The regression line has a positive slope. This means that these data show a positive relationship between the two variables in question, namely that Marsha's test scores tend to be higher for modules where she has studied more hours. However, there is no proof from this that the one has caused the other – i.e. that extra hours of study led to the higher marks. For example, it is possible that Marsha chose to work harder at the modules she enjoyed and was already good at and neglected her weaker modules. Proving causation is hard and this important question will be looked at more fully in the next chapter.

▶ Exercise 11.6

a The scattergraph and corresponding best-fit line are shown in Figure 11.11. The trend shows a positive relationship between age and pulse rate, i.e., if the evidence of this sample is typical of the adult population as a whole, then, in general, the older you are the higher your pulse rate.

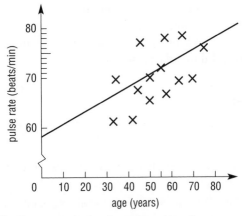

Figure 11.11 Scattergraph showing pulse rate and age

b The linear regression line calculated from these points is:

$$Y = 58.24 + 0.23X$$

where Y is pulse rate in beats per minute and X is age in years.

c Using the best-fit line calculated in (b), these values can be estimated by reading them off the graph, as follows.

To estimate Mia's pulse rate, start from her age (60) on the X axis, draw a line vertically up to the best-fit line and finally draw across to the Y axis. This gives an estimate of Mia's pulse rate of around 72 beats per minute. Or simply replace X by 60 in the equation of the time calculated in (b).

Mayerling's estimated pulse rate, by the same method, is around 59 beats per minute.

To estimate Tom's age, the same process is done in reverse. Start with the position on the Y axis which corresponds to Tom's pulse rate – 90 beats per minute – draw a line horizontally across until it meets the best-fit line and then, from the X axis, read off the age which this corresponds to. In this case, the estimate of Tom's age = 138 years! Or, you could replace Y by 90 in the equation of the line and work out the corresponding value of X.

d We have made a dangerous assumption with these estimates, namely that the same linear trend will continue in both directions beyond the range of values in the sample. In fact, this is not valid in either case here.

Key ideas

▶ This chapter dealt with how we might explore patterns in 'paired data'.

▶ Visually, the easiest way to see relationships between two variables is to plot the data onto a scattergraph.

▶ If the pattern of points appears to lie on a straight line, a useful way of summarizing the relationship is to find the equation of the 'line of best fit', or regression line, which can be made to pass through the scatter of points.

▶ Where the regression line has a positive slope, we say that the trend is positive, i.e. as one increases the other increases. Negative slopes suggest negative (sometimes known as 'inverse') relationships.

▶ Regression lines are useful for making forecasts. Particular dangers are attendant where this involves extrapolation (extending the regression line beyond known data) since there is no guarantee that the observed pattern will continue to apply.

▶ Regression can be thought of as a way of summarizing the relationship between two variables by collapsing the scatter of points into a single line. As such, the regression line reveals nothing at all about how widely scattered the points were around it.

▶ What exactly does the degree of scatter tell us? This is an important question in statistics and is the central idea of the next chapter.

To find out more via a series of videos, please download our free app, Teach Yourself Library, *from the App Store or Google Play.*

12

Correlation: measuring the strength of a relationship

In this chapter you will learn:

▶ *how to interpret patterns on a scattergraph*

▶ *how to calculate the correlation coefficient*

▶ *how to interpret correlation for data that are ranked in order*

▶ *that when two measures are correlated, this does not prove that one is the cause of the other.*

A long drought in Damascus in 1933 occurred just after a national craze of playing with a yo-yo. In their wisdom, the city elders decided that the movement of the yo-yos was influencing the weather and promptly banned them. The next day ... it rained! Meanwhile, in Madrid, a bullfighter, on the day of a *corrida*, will always shave his face twice. This is based on the belief that fear makes the beard grow faster and no bullfighter wants to display fear in the ring.

Most of us have a few 'off the wall' beliefs of this nature. I know someone who refuses to travel by train because about ten years ago he thinks he caught a cold on one! Based, often, on a tiny sample of experience or a few dubious anecdotes, we may deduce a cause-and-effect relationship where none exists.

Nugget: letting it out

During electrical storms, my granny would always open the front and back doors. She explained to me that this was to encourage the electricity to pass straight through her home, leaving her and her possessions untouched. As she added, '... and I haven't been struck yet!'

This chapter builds on the ideas of regression, which were the subject of the previous chapter, and explores the degree of confidence with which we can state that the relationship under consideration really does exist. In other words, we are interested in knowing about the strength of the relationship. An intuitive sense of how strong a relationship is can be deduced just by looking at the data drawn in a scattergraph and this is where we start in the opening section of the chapter. The next two sections deal with formal measures of correlation, the product – moment correlation coefficient and the rank correlation coefficient. Finally we look at the difficult question of whether strong relationships are necessarily cause-and-effect ones.

Scatter

As has already been suggested in the previous chapter, the regression line is useful as a statement of the underlying trend but tells us nothing about how widely scattered the points are around it. The degree of scatter of the points, or the *correlation*, is a measure of the strength of association between two variables. For example, perfect positive correlation will look like Figure 12.1.

Figure 12.1   Scattergraph showing perfect positive correlation

Perfect negative correlation will result from data which are plotted as in Figure 12.2.

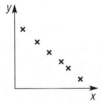

Figure 12.2   Scattergraph showing perfect negative correlation

In practice, most scattergraphs portray relationships with correlations somewhere between these two extremes. Some examples are shown in Figures 12.3 to 12.5.

Figure 12.3   Strong positive correlation

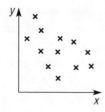

Figure 12.4 Weak negative correlation

Figure 12.5 Zero correlation

Exercise 12.1 Racing pulses?

The two scattergraphs shown in Figures 12.6 and 12.7 have been taken from Chapter 11 and both show the relationship between age and pulse rate but for two different age groups.

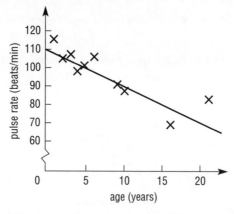

Figure 12.6 Scattergraph showing pulse rate and age of ten young people – Sample A

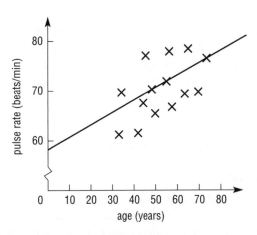

Figure 12.7 Scattergraph showing pulse rate and age of 14 adults –
Sample B

a How would you describe these relationships?
b How would you describe the strength of association in each case
and what might you deduce from this?

[Comments below]

a The scattergraph for Sample A shows a clear negative
relationship. Over the age range from 0 to 20 years, the pulse rate
typically drops from around 110 to about 70 beats per minute.
For the adults in Sample B, the relationship is a positive one, with
the pulse rate rising from around 65 for young adults to about 80
beats per minute for adults of around 70 to 80 years. We might
deduce from this that children's pulse rates decline quite a lot as
they get older, but with adults there is evidence of a slight increase
in pulse rates with age.

b Because the points on the scattergraph for Sample A lie fairly
close to the best-fit line, there seems to be a strong negative
correlation here. The corresponding pattern in Figure 12.7
suggests a rather weaker, but positive, correlation for Sample B.
What this implies is that you can expect to be able to make
predictions about the link between pulse rate and age with
greater confidence for children than for adults.

It is worth stressing that parts (a) and (b) in the previous exercise deal with different issues. Part (a) is concerned with describing *the basic underlying relationship* as represented by the best-fit line and is therefore a question about regression. Part (b) directs attention to *the strength of the relationship* as represented by the degree of scatter of the points around the regression line and is therefore all about correlation.

Product-moment correlation coefficient

Although it is useful to be able to gauge the strength of the relationship simply by looking at the scattergraph, a more formal method based on calculation is available which gives a numerical value for the degree of correlation. The product-moment coefficient of correlation (sometimes known as Pearson's coefficient) is calculated on the basis of how far each point lies from the 'best-fit' regression line. It is denoted by the symbol r.

The formula for r, the product-moment correlation coefficient, is:

$$r = \frac{S_{XY}}{S_X S_Y}$$

where X is the variable on the horizontal axis and Y is the variable on the vertical axis.

S_{XY} is the covariance, given by the formula $\dfrac{\Sigma XY}{n} - \bar{X}\bar{Y}$

S_X is the standard deviation of X $\sqrt{\dfrac{\Sigma X^2}{n} - \bar{X}^2}$ (1)

S_Y is the standard deviation of Y $\sqrt{\dfrac{\Sigma Y^2}{n} - \bar{Y}^2}$

The formula has been devised to ensure that the value of r will lie in the range -1 to 1. For example:

$r = -1$ means perfect negative correlation

$r = 1$ means perfect positive correlation

$r = 0$ means zero correlation

$r = -0.84$ means strong negative correlation

$r = 0.15$ means weak positive correlation

and so on.

Certain calculators with specialist statistical functions provide a key (usually marked 'r') for directly calculating the correlation coefficient. The exact procedure will be explained in the calculator manual but in principle you must first enter all the pairs of data values, then press the 'r' key. With only a conventional calculator to hand, however, calculating the value of r is really quite an effort. The following simple example sets out the three main steps in which the covariance, S_{XY}, the standard deviation of the X values, S_X, and the standard deviation of the Y values, S_Y, are first calculated.

A simple worked example for calculating r

Let us suppose that you have a simple data set consisting of only the following three pairs of values

X	Y
1	2
2	3
3	4

▶ **Step 1 Calculating the covariance, S_{XY}**

X	Y	XY
1	2	2
2	3	6
3	4	12
$\Sigma X = 6$	$\Sigma Y = 9$	$\Sigma XY = 20$

[1] This is a slightly different version of the formula for standard deviation given in Chapter 5, but they both amount to the same thing.

$$\overline{X} = \frac{6}{3} = 2 \quad \overline{Y} = \frac{9}{3} = 3$$

$$S_{XY} = \frac{20}{3} - 2 \times 3$$

$$= 6\frac{2}{3} - 6 = \frac{2}{3}$$

▶ **Step 2 Calculating the standard deviation of X, S_X**

X	X^2
1	1
2	4
3	9
$\Sigma X = 6$	$\Sigma X^2 = 14$

$$\overline{X} = 2$$

$$S_X = \sqrt{\frac{14}{3} - (2)^2}$$

$$= \sqrt{\frac{2}{3}}$$

▶ **Step 3 Calculating the standard deviation of Y, S_Y**

Y	Y^2
2	4
3	9
4	16
$\Sigma Y = 9$	$\Sigma Y^2 = 29$

$$\overline{Y} = 3$$

$$S_Y = \sqrt{\frac{29}{3} - (3)^2}$$

$$= \sqrt{\frac{2}{3}}$$

Finally, $r = \dfrac{S_{XY}}{S_X S_Y} = \dfrac{\dfrac{2}{3}}{\sqrt{\dfrac{2}{3}} \times \sqrt{\dfrac{2}{3}}} = \dfrac{\dfrac{2}{3}}{\dfrac{2}{3}} = 1$

So, in this example there is perfect positive correlation. When the three points are plotted on a scattergraph, this should make sense just by looking at their pattern. As you can see from Figure 12.8, the points form a perfect straight line with a positive slope.

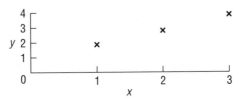

Figure 12.8 Data from simple worked example showing perfect positive correlation

Exercise 12.2 Another simple example showing the calculation of r

You can try an example on your own now. At this stage, you may still find it helpful to use simple artificial data just to clarify the main steps. For example, try the data below, which show perfect negative correlation, so you should expect to get a value for r of −1.

X	Y
1	4
2	3
3	2

Worked example

The next worked example is based on rather more interesting data. Notice how Table 12.1 provides you with an efficient, systematic way of calculating S_{XY}, S_X and S_Y. Here the calculation for r is based on the data given in Exercise 11.6.

Table 12.1 Pulse rate and age of 14 adults ($n = 14$) – Sample B

X	Y	X²	Y²	XY
(age)	(pulse rate)			
33	62	1089	3844	2046
50	66	etc.	etc.	etc.
45	77			
65	78			
57	78			
70	70			
42	62			
75	76			
57	67			
35	70			
44	68			
63	70			
50	70			
55	72			
$\sum X = 741$	$\sum Y = 986$	$\sum X^2 = 41\,281$	$\sum Y^2 = 69\,814$	$\sum XY = 52\,662$
$\overline{X} = 52.93$	$\overline{Y} = 70.43$			

$$S_{XY} = \frac{52\,662}{14} - 52.93 \times 70.43 = 33.71$$

$$S_X = \sqrt{\frac{41\,281}{14} - (52.93)^2} = 12.127$$

$$S_Y = \sqrt{\frac{69\,814}{2} - (70.43)^2} = 5.131$$

$$\text{Coefficient of correlation}, r = \frac{33.71}{12.127 \times 5.131} = 0.542$$

Clearly this calculation requires considerable effort, even with the aid of a calculator. If you need to find r repeatedly, it would be worth investing in a 'statistical' calculator which calculates it directly from the raw data. Alternatively, use a computer.

Exercise 12.3 Over to you!

In order to get practice at calculating the product-moment coefficient of correlation, have a go at some or all of the following and check that you come up with the same results as shown.

Table 12.2 The ages and pulse rates of a sample of ten young people – Sample A

Name	Age (years)	Pulse rate (beats per minute)
Irena	10	86
Nadia	3	107
Luke	4	98
Jon	6	105
Carrie	8	90
Theo	1	115
Jagjeet	16	69
Ira	21	81
Hannah	5	100
Alex	2	105

Source: personal survey.

[Comments follow]

The correlation coefficient for age against pulse rate for Sample A is -0.88. Intermediate values for calculating this value are given below.

$$S_{xy} = \frac{6553}{10} - 7.6 \times 95.6 = \text{-}71.26$$

$$S_x = \sqrt{\frac{952}{10} - (7.6)^2} = 6.12$$

$$S_y = \sqrt{\frac{93\,146}{10} - (95.6)^2} = 13.24$$

$$\text{Coefficient of correlation, } r = \frac{\text{-}71.26}{6.12 \times 13.24} = \text{-}0.88$$

Exercise 12.4 Interpreting *r*

How do you interpret the two values for *r* given in Exercise 12.3 and the worked example before it?

[Comments below]

The strength of correlation will be indicated by how far from zero the value of *r* lies. The Sample B value of $r = 0.542$ is reasonably strong, but the Sample A value of -0.88 is very high indeed, and suggests that among children there is a strong (negative) correlation between the two variables concerned. However, there are two important provisos which need to be considered when interpreting correlation.

The first point is the question of *batch size*. For Sample A, the batch size is 10 and for Sample B it is 14. In general, the smaller the batch size the larger the value of *r* must be in order to provide evidence of significant correlation. Thus, we must not be overly impressed with the Sample B value for *r* of -0.88 since the batch size is so small. This point, which relates to the degree of *significance* which we can attach to values of *r*, is explained more fully in the next chapter.

Second, a crucial issue here is being able to distinguish correlation between two variables and a *cause-and-effect relationship* between them. It is important to remember that correlation only suggests a pattern linking two sets of data. *Correlation does not prove that one of the variables has caused the other*. Indeed, exactly what the value of *r* does and does not tell us is quite difficult to interpret, and this issue is pursued in more detail in the final section of this chapter.

Rank correlation

It is not always necessary, or even possible, when investigating correlation, to draw on the sort of measured data that have been presented so far. An alternative is to work only from the rank positions. For example, when judging ice-skating or gymnastics competitions, the marks that are awarded are rather meaningless and their primary purpose is to allow the competitors to be ranked in order – first, second, third and so on. Information is sometimes stripped of the original data and the results presented only in order of rankings. For example, look at Table 12.3 which shows the UK top 10 boys' and girls' first names of 2015 published in *The Independent* newspaper. What, if anything, do these results suggest about the names that tend to get chosen?

Table 12.3 The UK top ten first names of 2015 along with their rankings for the previous year

Boys	This year	Last year		Girls	This year	Last year
Oliver	1	1		Amelia	1	1
Jack	2	2		Olivia	2	2
Harry	3	3		Isla	3	5
Jacob	4	4		Emily	4	3
Charlie	5	5		Poppy	5	7
Thomas	6	6		Ava	6	4
George	7	10		Isabella	7	8
Oscar	8	7		Jessica	8	6
James	9	9		Lily	9	12
William	10	8		Sophie	10	15

Source: *The Independent* newspaper, Monday 17 August 2015

There are some interesting investigations that can be made about the different patterns between girls' and boys' names. For example, notice here that all of the girls' names end in a vowel or 'y' as compared with only three boys' names. You may also have observed that the boys' list appears to be rather traditional, but quite how this characteristic could be measured and compared is not easy to see.

Nugget: you can learn a lot from a name

I carried out a similar investigation on boys' and girls' names with a group of students, who concluded that the choice of popular boys' names changes less than that of girls' names, year on year. When I asked why they thought this was, one said, 'We still live in a very traditional society where boys are taught to be the providers and girls are allowed to be artistic and creative. Successful providers have traditional names and successful "creatives" are allowed to have names that are a bit more fun.'

One property that we *can* test is the extent to which the two sets of names have retained their popularity since the previous year. A simple way to get a quick visual impression of this is to plot the rank positions of one year against the other on a scattergraph.

Exercise 12.5 Interpreting the scattergraphs

What do these two scattergraphs reveal?

[Comments follow]

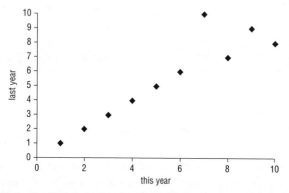

Figure 12.9 Scattergraph showing the ranks of boys' names against the previous year's position

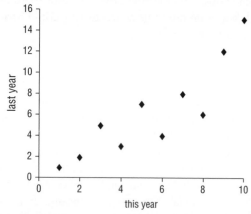

Figure 12.10 Scattergraph showing the ranks of girls' names against the previous year's position

A clear pattern in the boys' scattergraph suggests a positive correlation between the rank order of names in 2015 and that of the previous year. There is also evidence of a similar pattern in the girls' names, but the association is not quite so pronounced. This shows that the preferences for both boys' and girls' names changed very little over the year in question but there was slightly more change in the girls' names. But can we demonstrate this difference quantitatively?

The main theme of this section is to try to find a way of measuring the strength of association in these and similar scattergraphs where the data are in ranked form. While it would be possible to use the correlation coefficient, r, in such cases, there is an alternative measure which is specially designed for ranked data called the rank coefficient of correlation, r_s. It is sometimes known as 'Spearman's coefficient of rank correlation' after its inventor, Charles Spearman.

The formula for r_s is based on carrying out the following steps.

▶ calculate all the differences (d) between each pair of ranks,

▶ square the differences (d^2),

> Σd^2 is the sum of the squares of the differences in ranks

▶ add these squares (Σd^2),

▶ include this value in the formula below:

Coefficient of rank correlation, $r_s = 1 - \dfrac{6\Sigma d^2}{n(n^2 - 1)}$

Here is a worked example using the boys' names data from Table 12.4.

> n is the batch size

Table 12.4 Calculating r_s for the boys' names

Boys	This year	Last year	d	d²
Oliver	1	1	0	0
Jack	2	2	0	0
Harry	3	3	0	0
Jacob	4	4	0	0
Charlie	5	5	0	0
Thomas	6	6	0	0

(Continued)

Boys	This year	Last year	d	d²
George	7	10	-3	9
Oscar	8	7	1	1
James	9	9	0	0
William	10	8	2	4

Total = 14

$$r_s = 1 - \frac{6 \times 14}{10(99)} = 0.915$$

As with the product-moment correlation coefficient, r, the coefficient of rank correlation, r_s, has been devised to ensure that its value will lie within the range -1 to 1. The value for r_s can be interpreted in much the same way as the values for r were in the previous section. Thus:

$r_s = -1$ means perfect negative correlation

$r_s = 1$ means perfect positive correlation

$r_s = 0$ means zero correlation

$r_s = -0.84$ means strong negative correlation

$r_s = 0.15$ means weak positive correlation

and so on.

So clearly our value of 0.915 for the rank correlation coefficient for the boys' names shows a fairly high degree of correlation indeed. This is a statistical way of saying that the rank order of popularity of boys' names in 2015 was similar to the rank order of the previous year.

Exercise 12.6 Calculating r_s yourself

You will find it a useful exercise now to calculate a coefficient of rank correlation for yourself. Complete Table 12.5 and then work out r_s for the girls' names data from the table. How does the result compare with the corresponding figure for boys, and what does the comparison mean in terms of the question we have posed?

Table 12.5 Calculating r_s for the girls' names

Girls	This year	Last year	d	d²
Amelia	1	1		
Olivia	2	2		
Isla	3	5		
Emily	4	3		
Poppy	5	7		
Ava	6	4		
Isabella	7	8		
Jessica	8	6		
Lily	9	12		
Sophie	10	15		
				Total =

[Comments below]

You should have found that the value for r_s for the girls' names was -0.49. The calculation is set out in Table 12.6.

Table 12.6 Solution to Exercise 12.6

Girls	This year	Last year	d	d²
Amelia	1	1	0	0
Olivia	2	2	0	0
Isla	3	5	-2	4
Emily	4	3	1	1
Poppy	5	7	-2	4
Ava	6	4	2	4
Isabella	7	8	-1	1
Jessica	8	6	2	4
Lily	9	12	-3	9
Sophie	10	15	-5	25
				Total = 52

For the girls' names, $r_s = 1 - \dfrac{6 \times 52}{10\,(99)} = 0.685$

This value for r_s shows a strong positive correlation between the girls' names chosen in 2014 and 2015 but not quite as strong a correlation as there was for the boys' names (which had an r_s value of 0.915).

In this particular example, it is important to stress that we are identifying patterns in the top ten names over a particular two-year period in history within a particular country (the UK). These conclusions are not valid for other situations.

Cause and effect

> *Actually, the less I practise, the better I seem to get. The more I drink, the quicker my reactions are.*

For certain everyday events, the link between cause and effect seems fairly straightforward. We turn on a tap and water comes out; we fall over and it hurts! You would have to be a Cartesian philosopher to doubt the obvious connections between action and outcome in these examples. But not all relationships are so easy to interpret. Suppose, for example, that you suffer from persistent back pain and have tried a variety of remedies to cure it. You awaken one morning and think that it feels a bit better. Here is the sort of conversation you might be having with yourself.

> *Hmm ... it must be those exercises that I did last week that are beginning to have an effect. Perhaps I should do these exercises more regularly. Of course, it could be due to the restful bubble bath that I took last night. That's worth remembering [thinks 'I must buy another bottle of that bath oil...']. Hang on, though! On second thoughts, isn't there a bit of pain shooting down the old left leg? Yes, it's definitely nagging there. I must have overdone it on the exercise machine. I'd better lay off a bit in future...*

The problem with back pains and their remedies is that these things are extremely difficult to measure. Also, there are so many possible explanations for the various outcomes you can never be sure why they happened. As with many ailments, they may just get better over time anyway, despite your attempts at a drastic cure.

> *With proper medication the common cold usually lasts about a week but left to its own devices it can drag on for seven days.*

The relevance of these examples to correlation is to remind you that a strong correlation between two things does not prove that one has *caused* the other. A strong correlation merely indicates a statistical link, but there may be many reasons for this besides a cause-and-effect relationship. For example, traditionally the height of a pregnant woman and her shoe size were thought to be good predictors of whether or not she would need to give birth by Caesarean section operation. In general, it was thought women who were short in stature and petite of foot would also have narrow pelvises and would therefore find it difficult to have a normal vaginal delivery. And, indeed, the statistical evidence seems to bear this out, in that there has traditionally been a close correlation between women having a Caesarean birth and small feet. But this is not the same thing as saying that there is a close correlation between women *needing* a Caesarean birth and having small feet. It now seems that this is something of a myth which has grown over time. Doctors, believing the association to be valid, tend to prescribe Caesarean births for women with small feet, thereby making the correlation a self-fulfilling prophecy.

Such an association between two things is known as *spurious correlation* because it implies a cause-and-effect link which is not actually there. There are countless further instances of spurious correlation – that is where a high statistical correlation between two measures can give a misleading impression about the true nature of their relationship. For example, if you decided to correlate the national consumption of mozzarella cheese, over time, with the number of civil engineering doctorates awarded over the same period, you could well come up with a very high value for r. Similarly, it turns out that, over recent years, there has been a very close statistical correlation between the average consumption of margarine and the divorce rate. Common sense suggests that there is no clear cause-and-effect relationship with these two examples. The more likely explanation is that most things increase (or decrease) over time and charting the path of any two measures over time is likely to show a matching pattern.

Have a look now at Exercise 12.7, which shows three more examples of spurious correlation. You might like to entertain yourself by trying to think up some different possible explanations for the connections.

Exercise 12.7 Causal or casual?

Try to find at least two explanations for the following.

a The number of storks nesting in the chimneys of a sample of German villages is strongly correlated with the number of children in these villages.

b The number of cigarettes which people smoke is negatively correlated with their income.

c The number of crimes over time is positively correlated with the size of the police force.

[Comments below]

a *Babies and storks.* One possibility is that the storks are bringing the children to the villages. (Well, I suppose it is possible!) Another is that villages with more children tend to be larger and have more houses and therefore offer a larger number of chimneys for the storks to roost in.

b *Smoking and wealth.* Do people smoke because they are poor or are they poor because they smoke?

c *Police and crime.* Do we tend to employ greater numbers of police as a result of increases in crime, or do more police simply catch more criminals? A third possibility may be that a larger police force encourages a higher proportion of victims to report crimes (number of crimes committed is not the same thing as number of crimes reported). It is also likely that changes in patterns of crime over time may have less to do with the quality and quantity of the police officers we employ and much more to do with changes in the economic and value system of society as a whole.

These examples illustrate the difficulties in interpreting the result of being told that there is a strong correlation between two things. We really can't be sure which one was the cause and which the effect, or indeed whether both things were quite independent of each other and the changes were because of some other factor or factors.

Cause and effect can really only be tested under controlled conditions where all other possible influences are removed. Then one factor (the independent one) is systematically altered and the resulting change on the other one (the dependent factor) is observed.

Nugget: but how large a sample did you use?

It is important to stress here that the examples used in this chapter were all based on very small sample sizes (I've used sample sizes of as small as ten and seven). The reason for this choice was to make the calculations easier for the reader to follow. However, this may give a misleading impression about how big a sample size should be in order to provide convincing evidence about the strength of correlation. The truth is that, if you were to do some real research, you would need to use much larger sample sizes than in these illustrations.

In general, the larger your sample size the lower the value for r you need to come up with in order to believe that there was a strong positive correlation. For example, with a sample size of about 20, a reasonably convincing level of correlation could be taken to be around 0.6. Now increase the sample size to, say, 90 and you could have the same degree of confidence in the strength of the relationship based on a correlation coefficient value of only 0.3.

Politicians in office will point to a lowering of inflation and a rise in investment in the economy and expect to take all the credit, claiming that these were directly caused by their policies. Politicians in opposition will point to a lowering of educational standards and a rise in unemployment and expect the government to take the blame, claiming that these were directly caused by their policies. Political 'hawks' in the 1990s claimed that an easing in world tension (the ending of the 'Cold War') happened *because* they took a hard line and talked tough. The 'doves' argued that these changes happened *despite* the tough talking of the hard-liners and would all have taken place anyway for a set of quite different reasons.

Real-world judgements are not made in controlled laboratory conditions and the true nature of cause-and-effect relationships will always be open to many interpretations, subject to individual opinions, prejudices and vested interests.

Key ideas

▶ Correlation is a measure of the strength of association between two things. A useful intuitive indication of correlation is the extent of the scatter around the best-fit line when the data are plotted on a scattergraph. Positive correlation shows a clear pattern of points running from bottom left to top right, while negative correlation shows a pattern of points running from top left to bottom right.

▶ Two measures of correlation were introduced: the Pearson coefficient (r) and, for ranked data, the Spearman coefficient (r_s). Both measures produce values lying in the range -1 to 1. A coefficient value of -1 means perfect negative correlation, while a coefficient value of 1 means perfect positive correlation. When r or r_s take a value of around 0, there is no correlation between the two factors in question.

▶ Finally, caution was urged in the interpretation of correlation. Strong correlation is no indication of whether or not the relationship in question is one of cause and effect.

To find out more via a series of videos, please download our free app, Teach Yourself Library, *from the App Store or Google Play.*

13

Chance and probability

In this chapter you will learn:

▶ *ways of measuring chance such as odds and probability*

▶ *how to check your own intuitions about chance*

▶ *about adding and multiplying probabilities.*

Probability is a way of describing the likelihood of certain events taking place. Clearly different events have different likelihoods of occurring. At one extreme, some events may be thought of as being *certain* (the chance that you will eventually die, for example, is a certainty for mortal readers of this book), while at the other extreme another event might be described as *impossible* ('flying pigs', maybe). For most everyday situations, the degree of likelihood involved is neither certain nor impossible but will lie somewhere between these two extremes. You might like to spend a few minutes thinking about events whose outcomes are uncertain and some of the words we use to describe the degree of uncertainty.

You can talk about chance in terms of being 'very likely', 'fairly common', 'extremely unlikely', 'an even chance', 'a long shot', 'beyond all reasonable doubt' and so on. Although these terms are in common use, there is considerable empirical evidence to suggest that they don't mean the same thing to everyone. For example, some people would rate 'fairly common' as having a greater than even chance, while others would rate it much lower.

For most everyday purposes, these differences in meaning may not matter very much. However, there are other situations where words alone are just not sufficiently precise to describe and compare probabilities accurately.

Measuring probability

ODDS

Odds are the most common way of measuring uncertainty in situations where people are betting on an event whose outcome is unknown. In a horse race, for instance, the odds on each horse are a measure of how likely the bookmaker thinks each horse is to win a particular race. For example, if a horse is given odds of 5:1 (said as 'five to one' and usually written either as 5–1 or 5:1) this means that, given six chances, it would be expected to win once and lose the other five times. Suppose you bet on a horse with odds of 10:1 and the horse wins, then, for each £1 you bet, you win £10. If, instead, you had placed a bet of £5 you would win £50 as well as receiving the £5 that you had staked in the first place. If a horse is favourite to win, it is given short odds, perhaps 2:1 or 3:2, whereas unlikely winners will be given very long odds, perhaps 100:1. A near certainty might be described in betting jargon as an 'odds-on favourite'. This is where a horse is given such short odds that a winning bet will earn the punter less than their stake. Of course, they will get their stake back also. An example of an odds-on favourite might be odds of '2 to 1 on'. These are actually odds of '1 to 2'. In other words, betting £2 wins you a mere £1. The next exercise will give you a chance to calculate betting odds.

Exercise 13.1 Calculating odds

Below are the starting prices for an eight-horse race as they would appear in a newspaper. The betting is given at the bottom.

1   011   PLAY THE ACE (40) (E R Thomas) J Berry 9 7 . . . . .J Carroll 2
2   2525   AZUREUS (IRE) (16) (J C Murdoch) J S Wilson 9 5 . . . . .J Fanning (7) 5
3   414641   LAND SUN (IRE) (16) (John W Mitchell) M Channon 9 3 .C Rutter 8 V
4   1   COLWAY DOMINION (24) (K Stringer) J Watts 9 3 . . . . .Dean McKeown 7
5   153266   STAMFORD BRIDGE (17) (Mel Brittain) M Brittain 9 2 . . . . M Wigham 6
6   1   MARTINI EXECUTIVE (9) (D) (Martini Cars (Cheshunt Ltd) W Pearce 9 1 . . . . . . .D Nicholls 4
7   054   AMANDHLA (IRE) (75) (Neill Jackson) N Tinkler 9 0 . . . . K Darley 1
8   364   FYAS (25) (M H Easterby) M H Easterby 7 11 . . . . .L Charnock 3

— 8 declared —

BETTING: 9–4 Martini Executive, 3–1 Colway Dominion, 5–1 Land Sun, 7–1 Play the Ace, 10–1 Azureus, 12–1 Stamford Bridge, 14–1 Fyas, 16–1 Amandhla

Figure 13.1   The starting prices of an eight-horse race

Assuming you place a bet of £10 and ignoring the complications of tax, how much would you expect to win if you placed a winning bet on (a) the favourite, Martini Executive; (b) Amandhla; (c) Land Sun?

[Comments at the end of the chapter]

STATISTICAL PROBABILITY

In statistical work, probabilities are usually measured as numbers between zero and one and can be expressed either as fractions or as decimals. A probability of zero ($p = 0$) means zero likelihood, i.e. the outcome is impossible. A probability of one ($p = 1$) refers to an outcome of certainty. Usually, in statistical work, outcomes are the results of a *trial* or *experiment* ('tossing a coin', for example). Most trials have outcomes whose probabilities lie somewhere between zero and one. Thus, an outcome thought of as being fairly unlikely might have a probability of something such as 0.1, while an 'even chance' would correspond to a probability of 0.5, and so on.

Probability is about calculating or estimating the likelihood of various outcomes from particular trials. Clearly, if there is only one possible outcome from a particular trial, it is certain to happen and therefore its probability will equal 1. So, if we are involved in calculating probabilities, we must be dealing with trials with more than one possible outcome. Table 13.1 gives a few examples. As you look at it, focus on the distinction between a *trial* and the *outcome(s)* of that trial, as these two terms are often confused with each other. Understanding the distinction between these two terms will be particularly important when you study the sections describing mutually exclusive outcomes and independence later in this chapter.

Table 13.1 Events and outcomes

Trial	Possible outcomes	
Tossing a coin	1	Getting heads
	2	Getting tails
Choosing a playing card	1	Getting a spade
	2	Getting a heart
	3	Getting a diamond
	4	Getting a club
Having a baby	1	Getting a girl
	2	Getting a boy
Taking an examination	1	Passing
	2	Failing

You might like to add some of your own trials and their corresponding possible outcomes. With some trials, such as tossing a coin, there are only two possible outcomes.[1] With other trials, there are literally hundreds; for example, when someone puts their hand in a bag to select the winning raffle ticket. Of course, from the punter's point of view, there are only two outcomes of any importance here: winning or losing! Sometimes outcomes are all equally likely for a given trial. For example, if you select a card from a deck of 52 playing cards which has been thoroughly shuffled, it is equally likely that the card you choose is a spade, heart, diamond or club. The reason for this is that there are equal numbers of each suit – 13 in fact. With other trials, the various outcomes are not equally likely. For example, with most examinations, more people tend to pass than fail and therefore the probability of a randomly chosen individual passing is more likely than their failing. Similarly, slightly fewer girls are born than boys, so the probability of a baby being a girl is lower than being a boy (about 0.48 as against 0.52).

Nugget: trials, outcomes and events

In this chapter, I've tried to make a clear distinction between a trial and an outcome: a trial is like an experiment (such as rolling a die) which will generate an as-yet-unknown outcome. The trial 'rolling a die' has six possible outcomes, whereas 'tossing a coin' has two. Another word that often crops up in probability textbooks is 'event' and unfortunately this is a much misunderstood term. In everyday parlance, rolling a die might be described as an event. Unfortunately, this is not how the word is used in statistics. An event refers to an outcome or a combination of outcomes. For example, based on the trial 'rolling a die', here are some examples of possible events:

▶ getting a 4 (this is a single-outcome event)

▶ getting an even score (an event combining three outcomes)

▶ getting a score greater than 2 (an event combining four outcomes).

---

[1] There is a third outcome: that the coin lands on its edge. However, if the coin is tossed onto a smooth, hard surface, this outcome is so unlikely that its probability can be taken to equal zero.

When outcomes are equally likely it is fairly straightforward to calculate their probabilities. You simply divide the number of ways in which that outcome can occur by the total number of possibilities. Some examples are given in Table 13.2. Now do Exercise 13.2.

Exercise 13.2 Events, outcomes and probability

Table 13.2 shows how probabilities of outcomes are usually calculated. Check that you understand the first two and then copy the table and fill in the rest for yourself.

Table 13.2 Calculating probabilities

Trial	Outcome	Number of ways (n)	Number of possibilities (N)	Probability $\left(\frac{n}{N}\right)$
Tossing a coin	Getting 'tails'	1	2	$\frac{1}{2}$
Tossing a die	Getting a score of 5 or more	2	6	$\frac{2}{6} = \frac{1}{3}$
Choosing a playing card	The suit being 'hearts'			
Choosing a playing card	The card being an 'ace'			
A baby being born ...	... on a weekend			
Someone who was born in December ...	... having their birthday after Christmas Day			

[Comments at the end of the chapter]

In summary, a general definition of probability is $p = \dfrac{n}{N}$

where p is the probability of a particular outcome occurring, n is the number of ways that outcome can occur and N is the total number of possibilities.

It is important to remember that this definition of probability applies only in situations where the various ways in which the outcome in question can occur are equally likely. Outside the cosy world of perfectly shuffled packs of cards and perfectly symmetrical dice, this assumption is often unrealistic. Thus, to take the two examples from Exercise 13.2 of a December baby being born after Christmas Day and a baby being born on a weekend, at first sight it may seem reasonable to assume that babies are equally likely to be born on all days of the year. Yet, for hospital births, this may not be so. Many babies come into the world as a result of being medically induced and, given that hospitals may be short-staffed at weekends and over the Christmas period, inductions may not be so popular at this time. This is one reason why most textbooks stick to 'safe' examples involving dice, coins and cards when dealing with probability concepts.

Myths and misconceptions about probability

There are many examples of faulty intuitions in the area of probability. Indeed it is quite possible that you have some yourself. Exercise 13.3 will give you an opportunity to test out your intuitions in this area.

Exercise 13.3 Beliefs and intuitions

Which of the following do you think are true?

   i If you throw a six-faced die, a '6' is the most difficult outcome to get.

  ii If a coin is thrown six times, three heads and three tails is the most likely outcome.

 iii If a coin is thrown six times, of the two possible outcomes below, the second is more likely than the first (H = Heads, T = Tails)

   Outcome A   H H H H H H

   Outcome B   H T T H H T

  iv If two dice are thrown and the results added together, scores of 12 and 7 are equally likely.

   v If a fair coin was tossed 10 times and each time it showed 'heads' you would expect the eleventh toss to show 'tails'.

[Comments follow]

i Although many people believe that a 6 is the most difficult score to get when a die is tossed, this is actually false. There are several possible explanations for why this mistaken notion might persist. One may be that 6 is the biggest score, and therefore thought to be hardest to throw. A second may be that, in many dice games, 6 is the score that is required and therefore it seems logical that it should be the most difficult to get. But perhaps the most likely explanation is that people correctly feel that they are less likely to toss a 6 than to toss a 'not-6', and therefore it is more difficult to get a 6, than not to get a 6. However, this is quite a different belief from that of thinking that the six outcomes on a die are not equally likely.

ii This statement is correct. There are seven possible outcomes for this event, which range from getting no heads to getting six heads. The two least likely outcomes are the extremes of 'all six heads' and 'all six tails', whereas three heads and three tails is the most likely. The bar graph, Figure 13.2, summarizes the likelihoods of the various outcomes. Just why the graph looks like this and how to calculate the separate probabilities of each outcome (indicated on the vertical scale) are explained in the next chapter in the section called 'The binomial distribution'.

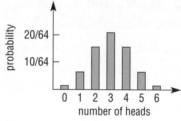

Figure 13.2 Bar chart of the likelihoods of the possible outcomes if a coin is thrown six times

iii This statement is false. Although an outcome with three heads and three tails is more likely than an outcome with six heads, the two sequences given here are in fact equally likely. The reason for this is that there is only one way of getting six heads in a row (HHHHHH), whereas there are many ways of getting three heads and three tails; for example, HTHTHT or HHHTTT or TTTHHH, etc. Taking order into account, all of these sequences are equally likely, so HHHHHH is as likely as HTTHHT. Again, this will be explained more fully in the next chapter.

iv This statement is false. There is only one way to throw a '12' (with a double six), whereas there are many ways in which a '7' can be obtained. The bar graph, Figure 13.3, shows the likelihood of each of the 11 outcomes (2, 3, ... 12) when two dice are thrown.

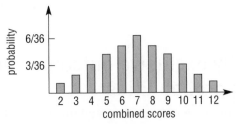

Figure 13.3 Bar chart of the likelihoods of the possible outcomes if two dice are thrown.

v The final statement is also false. Many people believe that, according to some vague sense of the ill-defined 'law of averages', an unexpectedly long run of heads will necessarily be corrected for by a tail on the next throw. Whether it is the coin itself or God who is responsible for this even-handed averaging out of outcomes is unclear. The reality is that the coin has no memory of past outcomes and begins each new toss afresh. This misconception is sometimes known as the 'gambler's fallacy' as it leads punters to believe that, after a string of losses, their next bet is more likely to win.

Nugget: the law of averages

The so-called law of averages contains the germ of a sensible idea, namely that, if you repeat a trial several times, the large and small outcomes will tend to 'even out' as a result of random variation. For example, suppose you wanted to know about the height of a particular flower and chose ten of them at random, the tall ones would tend to balance out the short ones so that you could use an average to find a middle value.

So far so good. But where many people's understanding of this law goes horribly wrong is when they assume that, in a small sample, the tall ones and the short ones *must* balance each other out. Random samples don't operate like that and this false belief is closely linked to the gambler's fallacy mentioned earlier. So common is this belief that when people begin a sentence with the words, 'Well, surely by the law of averages...', you can be pretty sure that you'll wish they hadn't completed the sentence!

Mutually exclusive outcomes and adding probabilities

When two (or more) outcomes are said to be 'mutually exclusive' this means that if one occurs the other cannot occur. For example, if a coin is tossed, the two possible outcomes 'getting heads' and 'getting tails' are mutually exclusive, because both cannot occur at the same time. Not all trials produce outcomes which are mutually exclusive. For example, when a baby is born, one outcome could be that it is a girl (rather than a boy). Another outcome might be that it has brown eyes (rather than blue or green). However, if a brown-eyed girl is born, both outcomes have occurred at the same time; clearly not mutually exclusive outcomes.

Exercise 13.4 Are the outcomes mutually exclusive?

Copy and complete the table below, indicating whether you think the outcomes listed for the particular events are mutually exclusive.

Table 13.3 Trials and outcomes

Trial	Outcomes	Outcomes mutually exclusive? y/n
(a) Tossing a coin	1 Getting heads	
	2 Getting tails	
(b) Rolling a die	1 Getting an even score	
	2 Getting an odd score	
(c) Rolling a die	1 Getting an even score	
	2 Getting a score greater than 3	
(d) Selecting a playing card	1 Getting an ace	
	2 Getting a spade	

[Comments at the end of the chapter]

Have a closer look at trial (c) in the table above. If the outcomes are represented as shown below, it is clear that 4 and 6 appear in both lists.

Even numbers	2	4	6
Numbers greater than 3	4	5	6

An alternative way of representing the outcomes of trials is by using rings corresponding to the different possible outcomes. The two outcomes being considered are shown here as two overlapping rings: outcome E (getting an even score) and outcome G (getting a score greater than 3).

all possible outcomes

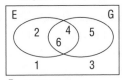

E = even score
G = greater score than 3

Figure 13.4 Rolling a die

When outcomes are not mutually exclusive, as is the case here, the rings will overlap. The region where they intersect corresponds to the particular scores (4 and 6) where both outcomes occur at the same time.

You may be wondering why it might be useful to know whether outcomes of certain trials are mutually exclusive. It is particularly important when *adding probabilities* to establish whether or not probabilities 'overlap', so as to avoid double-counting the region of intersection. A simple example based on outcomes E and G above should make this clearer.

The probability of outcome E can be written as $P(E)$. Similarly, the probability of outcome G can be written as $P(G)$.

Now $P(E) = 3/6 = 0.5$ (i.e. this is the chance of getting an even score)

and $P(G) = 3/6 = 0.5$ (i.e. this is the chance of getting a score greater than 3)

Supposing we were interested in finding the probability of either getting an even score or getting a score greater than 3, it would be tempting simply to add these two separate probabilities: $0.5 + 0.5$, giving the answer 1. However, this answer would seem to suggest that the probability of either one or other of E or G occurring is a certainty. This is clearly incorrect, because tossing a 1 or a 3 would not satisfy the condition. What has gone wrong here is that we have double counted the area of intersection shared by outcomes E and G. In fact, only four of the six possible outcomes satisfy either E or G, so the correct answer for the probability of either E or G occurring is 4/6.

Let us now look at how some of these ideas are expressed using more formal notation. In general, if a trial has several possible mutually exclusive outcomes (A, B, C, etc.) the probabilities are written as follows:

The probability of outcome A resulting is $P(A)$

The probability of outcome B resulting is $P(B)$

and so on.

If we want to find the probability of *either* A *or* B occurring, we *add* separate probabilities for A and B.

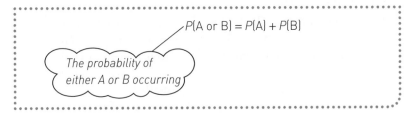

$P(A \text{ or } B) = P(A) + P(B)$

The probability of either A or B occurring

(In terms of the circle diagram in Figure 13.4, 'E or G' refers to the combined area covered by the two rings together.)

However, this rule only applies if outcomes A and B are mutually exclusive, otherwise the area of intersection has been double-counted.

If two outcomes are not mutually exclusive, the rule must be adjusted to take account of the double-counting, as follows.

For events that are not mutually exclusive:

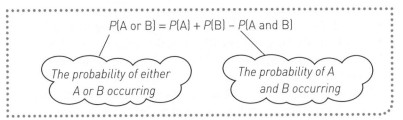

$P(A \text{ or } B) = P(A) + P(B) - P(A \text{ and } B)$

The probability of either A or B occurring

The probability of A and B occurring

(In terms of the circle diagram in Figure 13.4, 'E and G' refers to the area where the two rings overlap.)

When all the possible outcomes of a trial have been listed, the term to describe such a complete list is 'exhaustive'. For example, in Figure 13.4, the combined outcome 'E or G' is not exhaustive because outcomes 1 and 3 are not included. The possible outcomes 'heads' and 'tails' from tossing a coin are mutually exclusive (because if one occurs then the other cannot occur) and they are also exhaustive (together they cover all the possible outcomes).

Independence and multiplying probabilities

Two common terms which often get confused when dealing with probability are 'mutually exclusive' and 'independent'. Two outcomes are independent if the occurrence of one does not affect the likelihood of the other one occurring. What distinguishes 'independence' from 'mutually exclusive' is the sort of situation in which they tend to crop up. Talking about outcomes that are mutually exclusive is most usefully applied where the outcomes concerned arise from a single trial. On the other hand, the usefulness of the notion of independence is seen with reference to the outcomes of separate trials. For example, the need for introducing ideas of mutual exclusivity in Figure 13.4 arose from dealing with the problem of overlapping outcomes from a single trial – the tossing of a die. As will be shown in this section, however, independence is important when we wish to check whether the outcome of one trial is likely to affect the possible outcomes of some other trial.

A second distinction concerns the sort of calculation that will be involved. The idea of mutually exclusive outcomes tends to be an important consideration where you wish to know about the probability of *either* A *or* B occurring, and this is a situation where probabilities are being *added*. As you will see shortly, the idea of independence tends to be an important consideration where you wish to know about the probability of A *and* B occurring, and this is a situation where probabilities are being *multiplied*. Here is a simple example to illustrate why we multiply probabilities.

Example

Consider a game where first a coin and then a die are tossed in turn. You win a prize if the outcomes are 'heads' and '6', respectively. What is the probability of winning?

Solution

The probability of tossing 'heads' with a coin is $\frac{1}{2}$ and the probability of throwing a '6' is $\frac{1}{6}$. So the probability of being successful at both events is a half of $\frac{1}{6}$ or $\frac{1}{12}$. In terms of the original fractions, $\frac{1}{2}$ and $\frac{1}{6}$, what we have done here is to multiply them. Thus:

$$\frac{1}{2} \times \frac{1}{6} = \frac{1}{12}$$

So, if there are two separate outcomes, the method of calculating the probability of both occurring is to multiply the individual probabilities. It may seem slightly bizarre to multiply the probabilities when the question seems to be asking about taking two outcomes together (which may imply addition). This distinction between when to add and when to multiply probabilities when they are combined may not be immediately obvious. One simple rule of thumb is to think about whether the process of combining the probabilities has the effect of making the outcome *more* likely or *less* likely. Adding two numbers, even if each number lies between 0 and 1 (as all probability values do), will produce a larger result. Thus, since the chance of *either* A *or* B occurring is *more* likely than either one on their own, you must add $P(A)$ and $P(B)$. However, if you are considering the chance of A *and* B occurring, this is *less* likely than the chance of either A or B occurring on their own, so you must multiply $P(A)$ and $P(B)$. This can be summarized as follows.

Either A or B is ...	... more likely than each one separately ...	... so add the probabilities to get a bigger result
A and B is ...	... less likely than each one separately ...	... so multiply the probabilities to get a smaller result

We now turn to look specifically at the idea of independence, and the next exercise will get you thinking about what independence means. Do Exercise 13.5 now.

Exercise 13.5 Lottery

Many countries have a national lottery. One way in which this can be organized is that each week 'numbers' (perhaps available in the range between 1 and 999 inclusive) are sold to the public. Punters might pay something like £1 or £5 each for the number or numbers of their choice. The winning number is drawn at random and, after the lottery company has appropriated a suitable profit, the money remaining is split equally among the punters who bought that number.

Suppose you know that last week's winning number was 417, how might that information affect what number you might choose to buy this week?

[Comments below]

There are two separate elements in this choice. The first is to do with the laws of probability (specifically the notion of independence). The key question here is whether the draw for this week is independent of last week's outcome. Provided the draw is done fairly, then the events *should* produce independent outcomes, but you never know ... ! The second aspect to the question relates to the choices you might reasonably expect other people to make. In practice, most people don't believe in the independence of random events and would avoid choosing a number which won the previous week. (This is actually an example of the gambler's fallacy, described earlier.) So, given that all numbers have an equal chance of coming up, 417 would actually be the best choice since there are likely to be fewer people to share the prize with if you bet and won on that number.

Nugget: the Italian lottery agony

Nowhere in the world is the lottery more enthusiastically pursued than in Italy. On Wednesday, 9 February 2005, after not being drawn for nearly 2 years, the elusive number 53 finally showed up. Gripped by lottery fever and fuelled by media speculation and misinformation about the laws of probability, the longer the elusive 53 failed to show the more the punters bet on it. Unfortunately, many people started to bet money they didn't have, running up huge debts and having their homes repossessed. One distraught woman even drowned herself in the sea near Tuscany, leaving a note to her family stating that she had lost everything on the elusive number 53.

In total, during the months leading up to its appearance, more than €3.5bn (£3bn) was spent on '53', an average of €227 for each family.

Exercise 13.6 Tossing and turning

Suppose a fair coin was tossed ten times and each time it showed 'heads'. What would you expect the eleventh toss to show?

[Comments below]

This question was posed earlier in the chapter, but we can now take advantage of the language of 'independence' to explain the answer. Notice that the question states at the outset that the coin is 'fair'. This means that you can assume there is no bias either towards 'heads' or 'tails'. Also, under normal conditions, the successive tosses of a coin are independent events. In other words, the outcomes of the first ten tosses will not influence the outcome of the eleventh toss. Therefore, given that the coin is fair, the chance of heads or tails on the eleventh toss is still 50–50. Of course, if you didn't know that the coin was fair you might wish to argue that there was evidence of a bias towards 'heads', in which case you might feel that 'heads' was more likely on the next throw. However, there is never any argument for thinking that 'tails' is more likely after a run of heads and, interestingly, this is the most commonly held view.

Exercise 13.7 Lucky dip

A lucky dip consists of 20 envelopes, only 3 of which contain a prize.

a On the first dip, what is the probability of winning a prize?

b Suppose the first envelope chosen did contain a prize and the envelope is removed from the bag. What is the probability of the second dip also winning a prize?

[Comments below]

a Using the definition of probability given earlier, the probability of winning on the first dip is

$$\frac{\text{number of prizes}}{\text{number of possibilities}} = \frac{3}{20}$$

b After the first dip, the conditions for the second dip have been altered. In the latter case, the required probability is

$$\frac{2}{19}$$

In this instance, the second event (dip number 2) is *dependent* on the first (dip number 1).

A useful way of representing successive events is by using a tree diagram. Here are two examples: the first, Figure 13.5, uses a tree diagram to illustrate independent events and the second, Figure 13.6, uses a tree diagram to show dependent events.

As you can see from Figure 13.5, the probability of each outcome is written on the appropriate arrow. In order to find the probability of a particular outcome, say 'HH', simply trace the arrows from left to right and multiply the probabilities on the way.

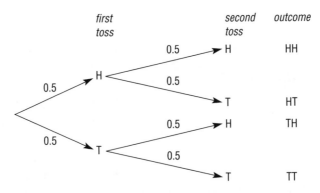

Figure 13.5 A tree diagram showing independent events

Exercise 13.8 Using the tree diagram to calculate probabilities with independent events

From Figure 13.5, calculate the probability that if two fair coins are tossed the outcome will be:

a two heads

b two tails

c one of each.

[Comments below]

a $P(\text{HH}) = \frac{1}{2} \times \frac{1}{2} = \frac{1}{4}$

b $P(\text{TT}) = \frac{1}{2} \times \frac{1}{2} = \frac{1}{4}$

c The probability of getting one of each = $P(\text{HT}) + P(\text{TH})$

$$= \frac{1}{2} \times \frac{1}{2} + \frac{1}{2} \times \frac{1}{2}$$

$$= \frac{1}{2}$$

A key feature of this example is that the probabilities on the arrows corresponding to each second toss are independent of the particular outcomes resulting from the first toss. Figure 13.6 illustrates events which are *not* independent. Note how this time the probabilities on the second set of arrows are *dependent* on the particular outcomes resulting from the first dip.

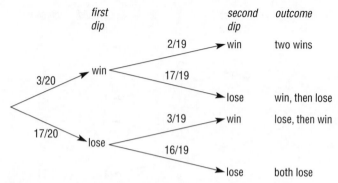

first dip | second dip | outcome

2/19 → win — two wins

3/20 → win

17/19 → lose — win, then lose

3/19 → win — lose, then win

17/20 → lose

16/19 → lose — both lose

Figure 13.6 A tree diagram showing dependent events

Exercise 13.9 Using the tree diagram to calculate probabilities with dependent events

Figure 13.6 represents the 'lucky dip' situation described in Exercise 13.7. Use the tree diagram to calculate the probability that, if two envelopes are chosen without replacement, the outcome will be:

a two wins

b no wins

c exactly one win.

[Comments below]

a $P(\text{WW}) = \dfrac{3}{20} \times \dfrac{2}{19} = 0.016$

b $P(\text{LL}) = \dfrac{17}{20} \times \dfrac{16}{19} = 0.716$

c The probability of getting exactly one win

$$= P(\text{WL}) + P(\text{LW})$$

$$= \dfrac{3}{20} \times \dfrac{17}{19} + \dfrac{17}{20} \times \dfrac{3}{19}$$

$$= 0.134 + 0.134 = 0.268$$

The importance of the idea of independence emerges when you are looking at the probabilities involved in a sequence of events – for example, the error factor in a series of runs of a particular machine.

Suppose, for the sake of argument, that the machine in question made identical components and had a chance of 1 in 100 of producing a faulty component on a particular production run. What is the chance of it producing a faulty component on each of three consecutive runs? In theory, you might be tempted to think that the probability is $\frac{1}{100}$ each time, so the overall probability is $\frac{1}{100} \times \frac{1}{100} \times \frac{1}{100}$ or 1 in a million. In practice, if a machine produces a fault, the fault tends to remain unless it is fixed. So, if the machine produces a faulty component on the first run, you might expect that, assuming it is not adjusted between runs, it is progressively *more* likely to behave badly on subsequent runs, giving a calculation of something like $\frac{1}{100} \times \frac{1}{20} \times \frac{1}{5}$. In other words, with machines, errors tend to cumulate and a series of production runs from the same machine will certainly *not* be independent of each other.

Just as the idea of mutually exclusive outcomes was shown to be important when adding probabilities, independence comes into its own when multiplying probabilities. Again, a few simple examples should help make the point.

Exercise 13.10 Three in a row!

a Suppose you toss a fair coin three times in succession. What is the probability of all three tosses coming up 'heads'?

b Suppose that you were able to take three dips in the lucky dip described in Exercise 13.7. What is the probability that you will win each time?

c A racing accumulator is the term used to describe a type of bet where you place a different bet on, say, three separate races and win only if all three horses come in first. If the odds on each of your chosen horses are 10–1, what are the odds of winning the accumulator?

[Comments follow]

a Since successive tosses of a fair coin are independent events, it is perfectly legitimate to multiply each of the separate probabilities involved. Thus:

Probability of three 'heads' in a row $= \dfrac{1}{2} \times \dfrac{1}{2} \times \dfrac{1}{2} = \dfrac{1}{8}$

b As was explained in the comments to Exercise 13.7, the probability of being successful in the second and third dips has altered as a result of the previous outcomes. Thus:

Probability of winning a prize in all three dips $= \dfrac{3}{20} \times \dfrac{2}{19} \times \dfrac{1}{18}$

$= \dfrac{6}{6840}$ (or 1 chance in 1140)

c In this case, it is reasonable to assume that the result of each race is independent of the others. Thus:

Probability of winning the accumulator $= \dfrac{1}{11} \times \dfrac{1}{11} \times \dfrac{1}{11} = \dfrac{1}{1331}$

In summary, then, if two trials are independent, the outcome of one does not affect the outcome of the other.

Comments on exercises

▶ **Exercise 13.1**

a £10 × 9/4 = £22.50

b £10 × 16 = £160

c £10 × 5 = £50

(plus, in each case, your £10 stake back)

▶ Exercise 13.2

Trial	Outcome	Number of ways (N)	Number of possibilities (N)	Probability $\left(\dfrac{n}{N}\right)$
Tossing a coin	Getting 'tails'	1	2	$\dfrac{1}{2}$
Tossing a die	Getting a score of 5 or more	2	6	$\dfrac{2}{6} = \dfrac{1}{3}$
Choosing a playing card	The suit being 'hearts'	13	52	$\dfrac{13}{52} = \dfrac{1}{4}$
Choosing a playing card	The card being an 'ace'	4	52	$\dfrac{4}{52} = \dfrac{1}{13}$
A baby being born...	... on a weekend	2	7	$\dfrac{2}{7}$
Someone who was born in December...	... having their birthday after Christmas Day	6	31	$\dfrac{6}{31}$

▶ Exercise 13.4

(a) Yes   (b) Yes   (c) No   (d) No

Key ideas

▶ Probability is an important topic in its own right. It allows us to understand, calculate and compare the risks around us.

▶ This chapter has focused on the basic notion of what probability means and how it is measured, using both odds and statistical probability. We have examined some common myths and misconceptions surrounding probability.

▶ There are some situations where probabilities are combined: situations of the form 'either or' require addition of the separate probabilities, while 'and' requires multiplication.

▶ There are two important ideas which can easily cause problems when probabilities are added and multiplied: the notion of mutually exclusive outcomes and independence.

▶ Probability theory also has a valuable role in helping us to draw sensible conclusions from data. In statistics, we attempt to look for patterns in data which show up both similarities and differences.

▶ A key role of probability is to help us decide how likely it is that any observed differences are simply due to natural variation. If it is very unlikely that the observed differences can be explained in this way, then we can deduce that the differences are statistically significant.

▶ In Chapter 14, Deciding on differences, you will see how these ideas of probability and data interpretation are brought together.

To find out more via a series of videos, please download our free app, Teach Yourself Library, *from the App Store or Google Play.*

14

Deciding on differences

In this chapter you will learn:

▶ *making comparisons in statistics*

▶ *how the laws of chance help us to make statistical decisions*

▶ *about the idea of a test of significance (otherwise known as hypothesis testing) and how it has parallels in legal decision making.*

In the last chapter, you looked at probability, how it is measured and what sort of meaning we can attach to adding and multiplying probabilities. All this may have seemed very far removed from your general view of what statistics is about. So, how do probability and statistics connect?

One of the aims of this chapter is to show how probability informs decision-making in statistics. As the chapter title suggests, one of the key decisions that has to be made in statistics is about differences. Here are a few practical examples:

▶ Does this growth hormone work?

▶ Does the weight of contents of this can of beans match what it says on the tin?

▶ Has the new traffic-calming measure improved traffic congestion?

▶ Is this drug more effective than that one?

All of these sorts of investigations involve measuring things and then making a comparison. At first sight, you might think that this should be a straightforward task. For example, taking the first two examples above:

Have the children taking the hormone shown a greater increase in their growth than those who didn't take the medication, or not?

Are the beans heavier than the stated weight, or not?

and so on.

Unfortunately, making these sorts of comparisons is a bit more complicated than that, and the reason is due to the fact of natural variation. For example, when children take a growth hormone, some may show a big growth spurt while others will show a rather small one. When you choose a sample of children, provide them with the medication and then calculate their mean growth increase, there is no way of knowing how representative this particular group of children is in terms of children's growth patterns overall. On the one hand, you might happen to choose a sample where the children showed an unrepresentatively small increase in growth, in which case the effectiveness of the hormone has been understated. On the other hand, if the sample happened to consist of children who showed an unrepresentatively large increase in growth, the effectiveness of the hormone would have been exaggerated.

It is down to the statistician to come to a sensible decision about any differences that can be observed, taking account of the fact that natural variation is going to muddy the waters. There are two key strategies that will help the statistician here. They are:

1 Choose samples that are as representative of the wider population as possible. Random sampling is a good way of achieving this.

2 Look at the amount of natural variation exhibited by the sample. This gives a good clue as to how much variation there is likely to be in the wider population and so this information will be useful when deciding on differences.

This area of statistics, where you are trying to decide on differences, is sometimes referred to as 'significance testing', also known as 'hypothesis testing'. Please note that this is considered to be an advanced topic in statistics. The aim in this chapter is to provide a very brief overview of this topic, so you should consult a more advanced textbook if you need to study it at a higher level.

Compared to what?

Here's a simple, medical illustration of how probability supports statistical decision-making. Suppose that you have succumbed to a fever with a high temperature. Your doctor prescribes a drug, which you take and, Hey Presto! the fever has gone next morning. How do you feel about the drug? You may feel sure that it did the trick ('It sure worked for me!'). However, it is possible that the fever had reached its peak by the time you got round to seeing the doctor, and you would have been well the next morning whether you had taken the drug or not. Looking at the wider picture, even if the drug did work for you, can you say how it might work for other people with a variety of similar symptoms resulting from subtly different causes? What exactly are you comparing it with?

Returning to the growth hormone investigation, let's suppose that the sample of children taking the hormone did show a large increase in growth. The statistician looking at this high sample mean value needs to ask how likely it would be to get a sample mean value as high as this from the wider population of children who had not taken the hormone. Only if this probability is very small can it be believed that the hormone has been effective.

So the key question here is the following:

> *How likely is it that we would get a result like this from chance alone?*

It is in the phrase 'by chance alone' that probability comes into the decision-making phase of any statistical investigation and this phrase is explored further in the next section.

By chance alone

If you are at all familiar with the world of espionage, you will know that special agent 007, James Bond, prefers his martinis to be 'shaken, not stirred'. This preference raises the interesting question as to whether Mr Bond could really tell the difference. Let's imagine an experiment in which he is given ten glasses of Martini, some of which have been shaken and some stirred. He is asked to sip each glass and come to a judgement about whether they are 'shaken' or 'stirred'. Then he's given a score showing the number of his successes. Judgements about how well Mr Bond has done will be based on comparing his score with what you might expect him to get from chance alone, which is a score of 5 out of 10, on average. With this in mind, let's set down some possible scores that he might get and consider your likely response to them.

Score	Response
0 – 5	Not very impressive Mr Bond. I could expect that you would get an average of 5 correct just by guessing, so this suggests that you cannot tell your 'shaken' from your 'stirred'.
10	Yes, a very good score. This might suggest that you really can tell the difference.
6 – 9	Hmmm. This is a grey area. Well done with coming up with a better score than what you would expect to get from guessing, but having said that, on a lucky day it still might be possible to get a score in this range just by guesswork.

This martini experiment is exactly equivalent to tossing a coin ten times and trying to guess the correct outcome each time. In statistics, this scenario is an example of what is known as a *binomial distribution* with ten trials ($n = 10$) and a 50:50 probability of success for each trial (i.e. p, the probability of success $= 0.5$).

The binomial distribution allows you to calculate the exact probabilities for getting 0, 1, 2, 3, …10 successes. The details of this calculation are beyond the scope of this book, but based on such calculations, it is possible to work out, for 100 trials of this experiment, the theoretical expected number corresponding to each outcome. These are set out in Figure 14.1. Note that to make the chart easier to read, the expected frequency of each outcome has been written above its corresponding bar in the bar chart.

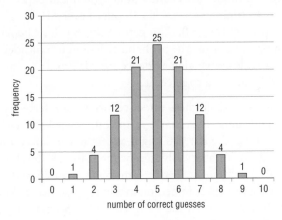

Figure 14.1 The theoretical number of correct guesses from 100 trials of the coin toss experiment

A test of significance

So how can we use information in the bar chart in Figure 14.1 to make sense of Mr Bond's results? Let's look at the following three possible scenarios.

Scenario A: he correctly identified 9 of the 10 martinis. This is equivalent to correctly guessing 9 or more of the outcomes of the ten coin tosses. By chance alone, you might expect this to happen only once (1 + 0, from the bar chart) out of every 100 trials. So there is a probability of 0.01 of getting this result or better by chance alone. This is such a tiny probability that we might be inclined to believe that Mr Bond really could distinguish his 'shaken' from his 'stirred' martinis.

Scenario B: he correctly identified 8 of the 10 martinis. This is equivalent to correctly guessing 8 or more of the outcomes of the ten coin tosses. By chance alone, you might expect this to happen 5 times (4 + 1 + 0) out of every 100 trials. So there is a probability of 0.05 of getting this result or better by chance alone. This is still a small probability and we might be inclined to believe in Mr Bond's claim – but not with such a strong degree of confidence as when he got 9 correct.

Scenario C: he correctly identified 7 of the 10 martinis. This is equivalent to correctly guessing 7 or more of the outcomes of the ten coin tosses. By chance alone, you might expect this to happen 17 times (12 + 4 + 1 + 0) out of every 100 trials. So there is a probability of 0.17 of getting this result or better by chance alone. This time the probability would be considered too large to provide convincing evidence for Mr Bond's claim and we would be inclined to reject it.

The decisions made in these three scenarios are referred to as tests of significance. In scenario A, the difference between his test score of 9 and what you might expect to get, on average, by guessing (5) is statistically significant at the 1% level. In scenario B, the difference between his test score of 8 and what you might expect to get, on average, by guessing (5) is statistically significant at the 5% level. But in Scenario C, the difference between his test score of 7 and what you might expect to get, on average, by guessing (5) is NOT statistically significant because the 17% probability of getting such an outcome by chance alone is too great.

A legal analogy

A helpful way of understanding the notion of 'test of significance' is to use the analogy of proof in the legal system.

INNOCENT UNTIL PROVEN GUILTY

In [English] law, we start with an important assumption – that the defendant is innocent until proven guilty. In deciding the innocence or guilt of the defendant, the burden of proof falls on the prosecution. In other words, it is the prosecution's job to provide sufficient evidence to show that the assumption of innocence doesn't stand up to scrutiny and therefore that the defendant is guilty 'beyond reasonable doubt'.

A similar perspective applies in statistics when a statistician looks at differences and tries to decide whether or not they are statistically significant. Taking the example of James Bond and his martinis, the statistician will start with the hypothesis that Mr Bond CANNOT distinguish 'shaken' and 'stirred' martinis. This assumption of 'no difference' is called the 'null hypothesis' and written as H_0 (said as 'H nought'). There should also be a statement of the alternative hypothesis – written as H_1 and said as 'H one'. In this example, the alternative hypothesis would be that Mr Bond CAN tell the difference between 'shaken' and 'stirred' martinis. In order to prove that Mr Bond can distinguish the drinks, the statistician will need evidence from the data to show that his null hypothesis is very unlikely, in which case it would be rejected in favour of the alternative hypothesis.

THE EVIDENCE

In a court case, the legal evidence for the prosecution will usually consist of a collection of events and information that shows the assumption of innocence to be unlikely. The better the prosecution's evidence, the less likely it is that the assumption of evidence will stand up to scrutiny. In the case of Mr Bond, the statistician will start with the question of what score would one expect him to get by guessing alone – on average, this would be a score of 5 out of 10. But if his actual score is, say, 8, this is so much greater than 5 that there would be only a 5% chance that the null hypothesis is true. As was indicated in the previous section, here we would say that the difference (between his actual and his expected scores) is significant at the 5% level. So, we can reject the null hypothesis in favour of the alternative hypothesis. And in the case where he scored 9 out of 10, this is an even more significant result. Here, the difference is significant at the 1% level.

These ideas are summarized in the table below.

Legal proof	Statistical proof
1. Start with the assumption of innocence of the accused.	1. Set up a starting hypothesis (referred to as the 'Null hypothesis' and written H_0) that Mr Bond cannot tell the difference, so his score, on average, should be 5. Also, set up an alternative hypothesis (H_1) stating that Mr Bond can tell the difference, in which case his score would be bigger than 5.
2. It is the job of the prosecution to provide evidence that shows the assumption of innocence to be unlikely. A guilty verdict should be reached only when the assumption of innocence fails to stand up BEYOND REASONABLE DOUBT.	2. It is the job of the statistician to provide evidence to discover whether the null hypothesis is unlikely and that there is indeed a difference in the data. A difference can be considered to be significant if there is only a tiny chance that such a difference could have occurred BY CHANCE ALONE. In these circumstances, the statistician would reject the null hypothesis (NO difference) in favour of the alternative hypothesis (yes, there IS a significant difference).

What now?

This chapter has provided only a very brief introduction to a large topic in statistics known as 'significance testing' or 'hypothesis testing'. As was indicated at the start of the chapter, if you wish to explore this topic more fully, you will need to consult a more advanced textbook and some good choices are suggested in the 'Taking it further' section at the end of the book. To support your further reading, the chapter ends with a glossary of some of the terminology that you have already met or are likely to meet in the future in your study of hypothesis testing.

Glossary of useful terminology in hypothesis testing

bias (in sampling)
Choosing an unrepresentative sample, which is likely to result in drawing a misleading conclusion about the wider population from which the sample was taken.

binomial distribution
The expected outcomes of a series of trials where each trial can result in one of just two possible outcomes (this is the 'two' that corresponds to the 'bi-' in binomial). These two outcomes may, or may not, be equally likely. An equally likely example is the repeated tossing of a fair coin, with outcomes 'heads' and 'tails'. A non-equally likely example is the repeated rolling of a die where the two outcomes being considered are 'getting 6' and 'getting not-6'.

the central limit theorem
This theorem is easiest to explain with a simple example. The standard measure of intelligence is called the Intelligence Quotient (IQ) and is defined so that, for the whole (adult) population, mean score = 100 and standard deviation score = 15. Imagine selecting one person at and time and measuring their IQ scores. Typical values will be 91, 103, 117, 89, and so on – quite widely spread around the overall mean of 100. Now imagine choosing random samples of, say, ten people at a time and finding the *mean* intelligence of these ten people. These mean scores will tend to cluster *much more closely* to the population mean of 100 (typical values might be 98.1, 103.2, 101.7, 99.2, and so on). The Central Limit theorem makes two claims in relation to situations such as this where repeated samples are taken from a population:

(a) the sample means will tend to cluster more closely around the population mean than would the individual values (indeed, the larger the sample size, the more marked this clustering will be);

(b) regardless of the underlying pattern of the original data from which the samples were taken, the set of sample means formed by repeated sampling will tend to conform to a particular pattern known as a normal distribution.

distribution

The distribution of a variable describes where the possible values are concentrated. More specifically, it shows the relative frequencies (or relative density) of the various possible values that the variable can take. For example, the expected distribution resulting from rolling a fair die many times will be uniform because each of the six outcomes (1, 2, 3, 4, 5 and 6) is equally likely. However, if a fair die is rolled, say, only 30 times, there is likely to be considerable variation in the distribution, due to natural variation. One of the best-known distributions is the normal curve, which is a useful model for describing many patterns in nature.

normal distribution

This is a particular pattern of data that is commonly found in nature where the middle or average values tend to predominate and the extremes (i.e. the very large and very small values) are less common. When graphed, this tends to produce a characteristic bell-shaped curve. It enables you to estimate the probability that the value of a randomly chosen item lies within a particular range.

null and alternative hypotheses (H_0 and H_1)

When carrying out a statistical test to determine whether two things differ significantly (say, whether girls tend to perform better than boys in speaking French) the test is usually set up in a formal manner, based on the starting assumption that there is *no difference* between girls' and boys' performance. This initial assumption of no difference is called the null hypothesis (null means 'amounting to nothing'). The alternative hypothesis is that the girls are better than the boys. The test is structured so that the burden of proof lies with being required to dismiss the null hypothesis. If you can't disprove the null hypothesis, you must then accept that it is true and the alternative hypothesis is not. If you can show that, based on the evidence of the data, the null hypothesis cannot be true beyond reasonable doubt, then, and only then, can you reject it in favour of the alternative hypothesis.

population
The entirety of those things under consideration. A common procedure in statistics is to take a sample from a population and use information from the sample to draw conclusions about the wider population. When the entire population is 'sampled', it is known as a census.

random sample
A sample that is selected so that each item has an equal chance of being selected.

sample
A selection of items taken from a wider population that is used to draw conclusions about the population from which it was taken.

statistical test
A formal procedure for determining whether two sets of data are significantly different from each other or whether one set of data is significantly different from a fixed value. Examples of statistical tests include the Z test, Student's t test and the Chi-squared test.

variability/variation
The key element that characterizes statistics, namely that data drawn from the real world have a tendency to vary. For example, if you visit an orchard and collect a basket of apples, they will not all have the same size or weight. Furthermore, not only will the apples in the basket vary naturally, but further variation will be imposed on the data due to inaccuracies and errors at the measurement and recording stages.

Key ideas

▶ This chapter has provided a whistle-stop tour of a major area of statistics called tests of significance, otherwise known as hypothesis testing. A key idea here is that, due to natural variation, you would expect differences to occur by chance. The essential principle underlying significance testing is that the difference being investigated must be sufficiently large to warrant reaching the conclusion that there is a 'real' difference, and unlikely to be one due to chance alone. It was suggested that a helpful way of understanding the notion of 'test of significance' is to use the analogy of proof in the legal system.

▶ Should you wish to read more about this topic, it was suggested that you consult a more advanced textbook and some choices are provided in the 'Taking it further' section of the book. A short glossary was also provided, which explained some of the terminology linked to hypothesis testing.

To find out more via a series of videos, please download our free app, Teach Yourself Library, from the App Store or Google Play.

Taking it further

General Statistics Background (from the author's bookshelf)

Dilnot, A. & Blastland, M. (2008), *The Tiger That Isn't: Seeing Through a World of Numbers*, Profile Books

Dubner, S. & Levitt, S. (2007), *Freakonomics: A Rogue Economist Explores the Hidden Side of Everything*, Penguin

Gladwell, M. (2009), *Outliers: The Story of Success*, Penguin

Goldacre, B. (2009), *Bad Science*, Harper Perennial

Huff, D. (1991), *How to Lie with Statistics*, Penguin Business

McCandless, D. (2012), *Information is Beautiful*, Collins

Paulos, J. A. (2015), *A Numerate Life: A Mathematician Explores the Vagaries of Life, His Own and Probably Yours*, Prometheus Books

Paulos, J. A. (1988), *Innumeracy: Mathematical Illiteracy and Its Consequences*, Viking Penguin

Pledger K. et al (2008), *Edexcel AS and A Level Modular Mathematics - Statistics*, Heinemann

Salkind, N. J. (2004), *Statistics For People Who (Think They) Hate Statistics*, Sage Publications

Senn, S. (2003), *Dicing with Death: Chance, Risk and Health*, Cambridge University Press

Sutherland, S. (2013), *Irrationality: the Enemy Within*, Pinter & Martin

Wiseman, R. (2004), *The Luck Factor: The Scientific Study of the Lucky Mind*, Arrow Books

More advanced statistics

Graham, A. et al (2004), *MEI Statistics 1*, Hodder Education

Graham, A. & Eccles, A. (2005), *MEI Statistics 2 (MEI Structured Mathematics (A+AS Level))*, Hodder Education

Jones, S. (2010), *Statistics in Psychology: Explanations without Equations*, Palgrave Macmillan

Rumsey, D. J. (2016), *Statistics for Dummies*, John Wiley & Sons

Rumsey, D. J. (2009), *Statistics II for Dummies*, John Wiley & Sons

Useful sources of data

A key source of data for the UK is the Office for National Statistics (ONS)

https://www.ons.gov.uk

As well as being responsible for collecting and presenting the ten-yearly UK census data, the ONS provides information on many areas of the lives of the UK citizen, organized under the following headings:

> Business, industry & trade;
>
> Economy;
>
> Employment and labour market;
>
> People, population and community.

British Social Attitudes

www.bsa.natcen.ac.uk

Useful websites

Maths is Fun provides many well-designed resources for mathematics learners:

www.mathsisfun.com

Robert Niles – Statistics that every writer should know:

www.robertniles.com/stats/

David Lane – an introductory statistics book online:
http://davidmlane.com/hyperstat/

Misuse and misconception of statistical facts:
www.cut-the-knot.com/do_you_know/misuse.shtml

Maths and statistics resources (some free):
www.mathgoodies.com

CAST (Computer-Assisted Statistics Teaching), a web-based
approach to teaching statistics using minimal mathematics and
based on using data:
http://cast.massey.ac.nz

The US National Library of Virtual Manipulatives (NLVM),
which provides many free maths and statistics applets:
http://nlvm.usu.edu/en/nav/vlibrary.html

The English National Curriculum online, Mathematics:
https://www.gov.uk/government/publications/national-
curriculum-in-england-mathematics-programmes-of-study

The author's personal website:
www.mathinpictures.co.uk

NRICH, which is linked to Cambridge University, offers a wide
range of excellent resources to support mathematical learning.
www.nrich.maths.org

Data and Story Library (DASL) – a US online library of data
files and stories that illustrate the use of basic statistics methods.
http://lib.stat.cmu.edu/DASL/

Other books written by Alan Graham

Graham, A. (2006), *Developing Thinking in Statistics*, Paul
Chapman Educational Publishing

Graham, A. (2017), *Teach Yourself Basic Maths*, Hodder
Education

Graham, A. (2008), *Teach Yourself Your Evening Class: Improve Your Maths*, Hodder Education

Graham, A. (2011), *Teach Yourself Basic Maths*, Hodder Education

Graham, A. (2011), *Statistics Made Easy (Flash)*, Hodder Education

Graham, A. (2011), *Statistics (Bullet Guides)*, Hodder Education

Graham, A. (2011), *Mathematics Made Easy (Flash)*, Hodder Education

Graham, A. (2011), *The Sum of You*, Hodder Education

Graham A. et al (2004), *MEI Statistics 1*, Hodder Education

Graham A. & Eccles A. (2005), *MEI Statistics 2 (MEI Structured Mathematics (A+AS Level))*, Hodder Education

Graham, A. & Duke, R. (2014), *Simple Statistics in Pictures*, iBookstore https://itunes.apple.com/gb/book/simple-statistics-in-pictures/id804004011?mt=11

Appendix

Choosing the right statistical technique

The flow chart below summarizes the main stages in statistical work. Relevant chapter numbers are given in brackets.

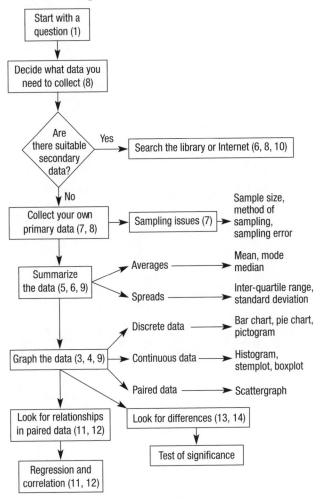

Index